# 15 Fundamentals for Higher Performance in Software Development

Includes discussions on CMMI, Lean Six Sigma, Agile and SEMAT's Essence Framework

Paul E. McMahon

# 15 Fundamentals for Higher Performance in Software Development

Includes discussions on CMMI, Lean Six Sigma, Agile and SEMAT's Essence Framework

Paul E. McMahon

This book is for sale at http://leanpub.com/15fundamentals

This version was published on 2014-07-08

ISBN 978-0-9904508-3-2

This is a Leanpub book. Leanpub empowers authors and publishers with the Lean Publishing process. Lean Publishing is the act of publishing an in-progress ebook using lightweight tools and many iterations to get reader feedback, pivot until you have the right book and build traction once you do.

©2014 Paul E. McMahon

# Tweet This Book!

Please help Paul E. McMahon by spreading the word about this book on Twitter!

The suggested hashtag for this book is #15fundamentals.

Find out what other people are saying about the book by clicking on this link to search for this hashtag on Twitter:

https://twitter.com/search?q=#15fundamentals

# Contents

Praise for 15 Fundamentals.... . . . . . . . . . . . . . . . . . . . . . . . i

Dedication . . . . . . . . . . . . . . . . . . . . . . . . . . . . . . . . . . . v

Foreword by Scott Ambler . . . . . . . . . . . . . . . . . . . . . . . . . vii

Acknowledgements . . . . . . . . . . . . . . . . . . . . . . . . . . . . . . xi

Introduction . . . . . . . . . . . . . . . . . . . . . . . . . . . . . . . . . . 1
   The Problem And What This Book Is About . . . . . . . . . . . . . . . . 1
   My First Goal in this Book . . . . . . . . . . . . . . . . . . . . . . . . 1
   About the Proposed Solution . . . . . . . . . . . . . . . . . . . . . . . 3
   How Performance (Process) Improvement Works Today . . . . . . . . . 3
   A Second Goal in Writing This Book and Why You Should Care . . . . . 4
   The Approach in this Book . . . . . . . . . . . . . . . . . . . . . . . . . 5
   A Third Goal of this Book . . . . . . . . . . . . . . . . . . . . . . . . . 6
   Summarizing the 15 Fundamentals and Thinking Framework Needs . . . . 6
   Neuroscience, Human Decision-Making and the Stories in this Book . . . 12
   This Book is for You if... . . . . . . . . . . . . . . . . . . . . . . . . . 12
   About the Terms Process and Practice in This Book . . . . . . . . . . . 13

## Part 1 – The Problem & The Start of a Solution . . . . . . . . . 15

Chapter One – The Sustainable Performance Dilemma . . . . . . . . . . . 17
   What Is Practice? . . . . . . . . . . . . . . . . . . . . . . . . . . . . . 17
   Looking Back At Yesterday And At Business Today . . . . . . . . . . . 19
   You Can't Sustain Performance Staying Where You Are . . . . . . . . . 20

# CONTENTS

|  |  |
|---|---|
| Patterns During Difficult Times | 21 |
| The View of Practice in Many Organizations Today | 22 |
| Sustainment Training | 25 |
| Chapter One Summary Key Points | 27 |

### Chapter Two – Repeating Specific Weaknesses/ Why You Should Care .. 29
| | |
|---|---|
| Background for Personal Improvement Project | 30 |
| Business Approach to Performance Improvement | 30 |
| Comparing Repeating Specific Weaknesses to Normal Process Variation | 31 |
| More About the Personal Improvement Project | 31 |
| A Business Example of a Repeating Specific Weakness | 33 |
| Repeating Specific Weaknesses: Critical to Sustainable Performance | 34 |
| Fundamentals First, But Why They Aren't Enough | 35 |
| Keeping People Motivated Throughout Their Careers | 37 |
| Chapter Two Summary Key Points | 39 |

### Chapter Three – Common Measurement Mistakes . . . . . . . . . . . . . . 41
| | |
|---|---|
| Common Organizational Measurement Mistakes | 41 |
| Start With Your Current "As-is" Process | 42 |
| Where to Look for Your Most Valuable Repeating Specific Weaknesses | 44 |
| A Case Study of a Common Measurement Mistake | 45 |
| A Flaw Observed With Today's Common Analysis Approach | 48 |
| Where to Look for Candidate Areas to Measure | 49 |
| Chapter Three Summary Key Points | 50 |

### Chapter Four – Little Things That Aren't So Little . . . . . . . . . . . . . . . 51
| | |
|---|---|
| Attaining High Value Improvements | 51 |
| Collect, Analyze, Act: Fundamental to Successful Improvement | 52 |
| Typical Business Stumbling Blocks | 52 |
| Business Case Study Demonstrating Sustained Performance | 53 |
| What's Happening Today in CMMI Level 5 Organizations? | 54 |
| Problem Isn't That We Aren't Trying | 54 |
| Gaining High Value Performance Benefits from Small Changes | 55 |
| A Small Change Requiring Practice To Master | 56 |
| Helping Practitioners With Small Changes | 57 |
| Another Reason to Focus on Mastering Small Changes | 57 |

    Personal Improvement Project Repeating Specific Weaknesses . . . . . . .   59
    Chapter Four Summary Key Points . . . . . . . . . . . . . . . . . . . . .   61

## Chapter Five – First Level Checkpoints: Necessary, But Not Sufficient . .   63
    How I Countered My Repeating Specific Weaknesses . . . . . . . . . . .   64
    Definition and Characteristics of First Level Checkpoints . . . . . . . . .   64
    Performance Objectives, Measures and First Level Checkpoints . . . . . .   65
    How First Level Checkpoints Work . . . . . . . . . . . . . . . . . . . .   66
    Why First Level Checkpoints Aren't Enough . . . . . . . . . . . . . . .   67
    First Level Checkpoints in the Business World: Quality Assurance . . . . .   68
    An Example of Improving Your Quality Assurance Checklists . . . . . . .   69
    Common Mistakes with First Level Checkpoints . . . . . . . . . . . . .   70
    Examples of 1st Level Checks/ Where Business Needs More Help . . . .   71
    Chapter Five Summary Key Points . . . . . . . . . . . . . . . . . . . .   73

# PART II Beyond Fundamentals . . . . . . . . . . . . . . . .   75

## Chapter Six: The Path to Higher Performance . . . . . . . . . . . . .   77
    Starting to Discover Better Performance Checklists . . . . . . . . . . . .   78
    The Wrong Way to Conduct Root Cause Analysis . . . . . . . . . . . . .   79
    Challenges Related to Effective Root Cause Analysis . . . . . . . . . . .   80
    Discovering Second Level "Feel" Checkpoints . . . . . . . . . . . . . .   82
    Motivating the Need for a Different Kind of Practice . . . . . . . . . . .   83
    Example of Second Level Checkpoint In Business: Big Picture Feel . . . .   84
    Chapter Six Summary Key Points . . . . . . . . . . . . . . . . . . . . .   87

## Chapter Seven: The Essentials of Second Level Checkpoints . . . . . . .   89
    Simplifying Complexity Through Patterns . . . . . . . . . . . . . . . .   90
    How Senior Management Can Help Practitioners Make Better Decisions .   91
    How Second Level Checkpoints Differ from First Level Checkpoints . . . .   92
    2nd Level Example: Requirements Incomplete or Ambiguous . . . . . . .   93
    Summarizing the Essentials of Second Level Checkpoints . . . . . . . . .   94
    Making Second Level Checkpoints Easier Through Patterns . . . . . . . .   95
    Chapter Seven Summary Key Points . . . . . . . . . . . . . . . . . . .   96

## Chapter Eight: Feeling Your Way to Higher Performance . . . . . . . .   97

# CONTENTS

    You Have to Get to Prevention for Higher Performance . . . . . . . . . . 97
    Sensing the Wrong "Can Do" Vision for High Value Payback . . . . . . . . 98
    An Example of "Feel" Leading To Better Decisions . . . . . . . . . . . . 99
    Feel Versus First Level Checkpoints . . . . . . . . . . . . . . . . . . . 100
    Practice to Help Keep Your Eye on Your Goal . . . . . . . . . . . . . . . 101
    Chapter Eight Summary Key Points . . . . . . . . . . . . . . . . . . . 104

**Chapter Nine—Better Decisions Through Better Practice With Patterns** . 105
    What Is Integrated Practice? . . . . . . . . . . . . . . . . . . . . . . 106
    What Is a Pattern? . . . . . . . . . . . . . . . . . . . . . . . . . . . 106
    The Good And The Bad Side of Patterns . . . . . . . . . . . . . . . . 106
    Creating the Right Patterns: Software Developer Examples . . . . . . . . 107
    Recalling the Procurement Case . . . . . . . . . . . . . . . . . . . . 108
    Spreading Positive Performance Across Your Organization . . . . . . . . 109
    Utilizing the "Listen-for" Technique For Positive Outcomes . . . . . . . 110
    Example Using Patterns to Aid Personal Improvement . . . . . . . . . 112
    Example Using Patterns To Improve Business Performance . . . . . . . 112
    What Does Practice Have To Do With Performance? . . . . . . . . . . 115
    Chapter Nine Summary Key Points . . . . . . . . . . . . . . . . . . . 116

**Chapter Ten—Right Patterns Through Practical High Maturity** . . . . . . 117
    Motivation for the Techniques Discussed in this Chapter . . . . . . . . 117
    In Search of the Real Root Cause on the Personal Improvement Project . . 118
    The Organizational Performance Impact of Inaccurate Data . . . . . . . 119
    Business Example: Impact of Taking Action with Inaccurate Data . . . . 120
    The Power of Small Changes . . . . . . . . . . . . . . . . . . . . . . 121
    Reinforcing An Important Insight About Small Changes . . . . . . . . 122
    Example of Small Change To Counter the Dictated Schedule Scenario . . . 123
    Helping Your People Make Effective and Timely Small Changes . . . . . 123
    Revisit the Common Measurement Mistake from Chapter Three . . . . . 124
    The Problem and How Lean Six Sigma Can Help . . . . . . . . . . . . 126
    Putting the Pieces Together With a Structured Real Story . . . . . . . . 127
    Implementing CMMI High Maturity Practices in a Practical Way . . . . 129
    Requirements Engineer's Story Voice Track . . . . . . . . . . . . . . . 130
    Key Characteristics of Structured Real Stories . . . . . . . . . . . . . . 132
    Increasing Involvement with Process Performance Baselines and Models . 133

    Business Example Motivating the Structured Real Story Technique . . . . 135
    Chapter Ten Summary Key Points . . . . . . . . . . . . . . . . . . . . 138

## Chapter Eleven– Practical High Maturity and Agile Retrospectives . . . . 139
    Statistical Process Control (SPC) Simplified . . . . . . . . . . . . . . . 140
    1st Example of SPC: The Frustrated Process Engineer . . . . . . . . . . 143
    2nd Example of SPC: Empowering Your Scrum Team . . . . . . . . . . 145
    How an Unstable Process Could Help Your Team's Performance . . . . . 148
    Chapter Eleven Summary Key Points . . . . . . . . . . . . . . . . . . 149

## Chapter Twelve– The Real Root Cause and Solution . . . . . . . . . . . 151
    The Real Root Cause on the Personal Improvement Project . . . . . . . . 152
    The Solution . . . . . . . . . . . . . . . . . . . . . . . . . . . . . . 153
    Insight Into Repeating Specific Weaknesses . . . . . . . . . . . . . . . 155
    Ask Why 5 Times to Guide Your Search for Root Causes . . . . . . . . . 156
    How Allowing Variation Can Reduce Variation Where it Counts . . . . . 157
    The Right Kind of Practice . . . . . . . . . . . . . . . . . . . . . . . 158
    Chapter Twelve Summary Key Points . . . . . . . . . . . . . . . . . . 159

# PART III A Thinking Framework . . . . . . . . . . . . . . . 161

## Chapter Thirteen: An Example–The Essence Thinking Framework . . . . 163
    Review of Problem we all Face . . . . . . . . . . . . . . . . . . . . . 164
    Why Are We Falling Short When Millions Are Spent Annually? . . . . . 165
    What's Really Wrong with How we Implement Performance Improvement? 166
    Understand the Problem, Carefully Approach the Solution . . . . . . . . 167
    Locating the Right Solution . . . . . . . . . . . . . . . . . . . . . . . 168
    Handling the More Difficult 2nd Type Situations . . . . . . . . . . . . 169
    The Essence Approach . . . . . . . . . . . . . . . . . . . . . . . . . 170
    The Essence Framework . . . . . . . . . . . . . . . . . . . . . . . . 171
    The Seven Alphas Inside the Essence Framework . . . . . . . . . . . . 171
    Essence Checklists Helping With Real Practitioner Pain Points . . . . . . 173
    How Esssence Checklists Differ From Traditional Checklists . . . . . . . 174
    Examples How Essence Checklists Go Beyond Existence Checklists . . . . 175
    Example 1: Checklist item Requirements Alpha Coherent State . . . . . 175
    Example 2: Checklist item Stakeholders Alpha In Agreement State . . . . 176

# CONTENTS

    Example 3: Checklist item Team Alpha Formed State . . . . . . . . . . . 176
    Example 4: Checklist item SW System Alpha Architecture Selected State . 177
    Not Checklists an External Auditor Can Easily Apply . . . . . . . . . . 177
    An Example of Using Essence as a Thinking Framework . . . . . . . . . 178
    Examples Essence Helping Detect Practitioner Practice Pain Points . . . . 178
    Helping Us Understand Practitioner Frustration . . . . . . . . . . . . . 179
    Example Essence Powering CMMI, Lean Six Sigma, Agile Retrospectives . 180
    Example of a Team Using Essence Independent of Practices . . . . . . . 181
    How Essence Relates to Two Types of Pain Points . . . . . . . . . . . . 183
    Chapter Thirteen Summary Key Points . . . . . . . . . . . . . . . . . . 184

**Chapter Fourteen: Better Decisions, Better Practice, and Essence Patterns**   185
    What is an Essence Pattern? . . . . . . . . . . . . . . . . . . . . . . . 185
    Pattern 1: The Dictated Schedule . . . . . . . . . . . . . . . . . . . . . 186
    Scenario Assessment and Consequence of Decision . . . . . . . . . . . 190
    A Decision that Could Have Improved the Team's Performance . . . . . 191
    Scenario Summary . . . . . . . . . . . . . . . . . . . . . . . . . . . . 192
    Pattern 2: A Little Coaching at Just the Right Time . . . . . . . . . . . 196
    Essence Thinking Framework Helping With Gentle Reminders . . . . . 197
    Apply the Right Practice at the Right Time With the Thinking Framework 200
    Pattern 3: Keeping an Opportunity Alive and Healthy . . . . . . . . . . 201
    Four Key Points from the Zack Example . . . . . . . . . . . . . . . . . 211
    Summarizing the Idea of Essence Patterns . . . . . . . . . . . . . . . . 212
    Chapter Fourteen Summary Key Points . . . . . . . . . . . . . . . . . 213

**Chapter Fifteen: Conclusion** . . . . . . . . . . . . . . . . . . . . . . . . . 215

**Epilogue** . . . . . . . . . . . . . . . . . . . . . . . . . . . . . . . . . . . 221

**Appendix A – Building a Library of Patterns** . . . . . . . . . . . . . . . . 227

**Appendix B – Performance Models Aiding Practitioner Daily Decisions** . 229

**Appendix C – Simple Example of Second Level Checkpoint** . . . . . . . . 231

**Appendix D – Comparing Performance Improvement Frameworks** . . . . 233

**Appendix E – More Evidence Supporting Need for Integrated Practice** . . 235

Appendix F – Cross-Reference CMMI Specific Topics in Book . . . . . . . 237

Appendix G – Cross-Reference Lean Six Sigma Specific Topics in Book . 239

Appendix H – Cross-Reference Agile Specific Topics in Book . . . . . . . 241

Appendix I – Cross-Reference Practitioner Scenarios . . . . . . . . . . . 243

Appendix J – Intentionally left blank . . . . . . . . . . . . . . . . . . . . 245

Appendix K – About the Term Pattern in this Book and An Analogy . . . 247

Appendix L – Why Risk Is Not an Essence Thinking Framework Alpha . 249

Appendix M – Questions About the Essence Framework . . . . . . . . . . 251

Appendix N – A Caution When Using the Essence Framework . . . . . . 253

Appendix O – Intentionally left blank . . . . . . . . . . . . . . . . . . . . 255

Appendix P – 6 Additional Essence Patterns . . . . . . . . . . . . . . . . 257
    Pattern 4: Reasoning Through a Requirements Dilemma . . . . . . . . 257
    Pattern 5: Reasoning Through a Lack of Data Challenge . . . . . . . . 262
    Pattern 6: Taking Responsibility for a Key Requirement . . . . . . . . 270
    Pattern 7: Using Data to Aid Practical Improvement . . . . . . . . . . 274
    Pattern 8: The Late Hardware Dilemma . . . . . . . . . . . . . . . . . 279
    Pattern 9: Assisting Self-Direction . . . . . . . . . . . . . . . . . . . . 282

Appendix Q – Practical High Maturity and Essence . . . . . . . . . . . . 287

Appendix R – Helping Your Team Achieve Higher Competency Faster . . 297

Appendix S – Cross Reference Critical Competency Needs, Case Studies 301

Appendix T – Which Organizations Should Care About Essence . . . . . 303

References . . . . . . . . . . . . . . . . . . . . . . . . . . . . . . . . . . . 305

About the Author . . . . . . . . . . . . . . . . . . . . . . . . . . . . . . . 311

# Praise for 15 Fundamentals...

"Paul McMahon has a refreshingly different approach to process improvement. It is based on sound theory, personal observations over a long career, and his participation in the groundbreaking SEMAT project. This book is a perfect counterpoint to the traditional improvement methods so often applied with CMMI, Six Sigma, and other improvement approaches. Read it and then wonder how you ever survived those other approaches."

*Dr. Richard Turner, Co-Author of CMMI Survival Guide: Just Enough Process Improvement*

"Many efforts have been made to capture the enormous breadth and depth of knowledge and practices necessary to improve organizational performance. Until now, these efforts have either woefully lacked for substance or completeness. More likely, other efforts simply gave up on the fantasy of being complete and settled for a focus on a tiny subset of what's needed to be said. Through masterful architecting of ideas, Paul's *15 Fundamentals for Higher Performance in Software Development* offers a readily digestible framework to provide both substance and completeness to a very large, complex and important subject."

*Hillel Glazer, Author, High Performance Operations, CMMI High Maturity Lead Appraiser*

"Paul has done a great job looking at a wide spectrum of prevalent software methods, without a bias for or against any, and come up with a few practical tips for a sustained performance improvement. If you are short of time or don't know where to start the book, jump to the 15 fundamentals. They are worth their weight in gold."

*Prabhakar R. Karve, Director of Engineering, Impetus*

"I very much like the 'out of the box' approach. The book exemplifies high maturity thinking in a simple way. There is a need for this book because people are always looking for examples of high maturity thinking."

*Winifred Menezes, CMMI High Maturity Lead Appraiser*

"This book gave me insight into other ways to improve and lay the foundation for CMMI Level 4 and 5; a foundation to last and expand."

*Dr. Michael Oakes, Process Improvement Lead, Alion Science and Technology, A CMMI Development and Services Level 3 Organization*

"I am one of those people that Paul refers to in the preface of this book who are turned off and have tuned out when it comes to the multitude of process and performance improvement approaches along with their related hype and buzzwords. So the first thing that hit me when Paul asked me to review this book was 'oh no! not another buzzword and tool!' But in going through it, it really hung together. This actually struck home with me because this Essence framework introduces a lot more context with regard to problem solving. And again this upset me because I wanted to scream at someone for introducing another tool and I couldn't."

*John Troy, Program Manager Rockwell-Collins*

"Congratulations on producing a really valuable piece of work. Clearly the product of many years' worth of serious application. The book reveals the essence of improving software development performance by teaching us to take ownership for improvement and focusing on patterns to address repeating, specific, weaknesses."

*Barry Myburgh, Johannesburg Centre for Software Engineering(JCSE), School of Electrical and Information Engineering, University of the Witwatersrand, Johannesburg, South Africa*

"We have spent about 100 years applying scientific management. Until now we just do our job without knowing why we are doing it. We know what we are doing when we pay attention but not when we don't pay attention and it is in those times that our efforts come undone. This is the thrust of a lot of this book."

*Dr. Tom McBride, University of Technology, Sydney, Australia*

"The material addresses how to apply principles and practices to control the Complex Adaptive System that we call our organization. SEMAT provides structure to control development and sufficient freedom to leverage the creativity from within the organization, to meet your objectives. If individuals take the time to "think" about how to improve and their team understands "why" they are performing an activity; the leader and the team should be able to adopt and extend the SEMAT framework to address their desired 'productivity' performance goals."

*Bob Epps, Lockheed Martin Corporate Engineering and Technology*

"If you are a software practitioner and serious golfer (or athlete), and are looking for advice on how to continuously improve your professional and personal performances, this book is for you. It encompasses 40 years of Paul's experience on how to practice to get to the next level of excellence, both at work and on the green."

*Dr. Cecile Peraire, Carnegie Mellon University, Silicon Valley Campus*

# Dedication

To my family for their continual support and encouragement; to the community of SEMAT volunteers for their support and insightful discussions; and to Watts Humphrey who was an inspiration to me in the final months of his life.

# Foreword by Scott Ambler

The goal of a foreword is to help you to determine whether or not you want to invest your time reading this book. Some people read forewords but I suspect many do not. You apparently are one of those few people who read forewords. Good for you.

A few years ago I saw Watts Humphrey speak about PSP, TSP, and CMM/CMMI at an agile conference. After his keynote we had a chance to break bread and discuss how CMM(I) had worked out in practice. Watts told me that he felt that the CMMI community had badly gone off the rails, given many pushes by large service providers who were primarily interested in getting as much money out of the US Federal government as they possibly could. Worse yet, these service providers were aided and abetted by a systemic culture within their customers that rewarded following procedures rather than doing what it takes to deliver real value to their stakeholders. Watts lamented that this was exactly opposite of what he had envisioned for CMM(I). He also recognized that the agile community not only understood what was really important they also had a viable strategy for achieving that goal. He saw the need for the CMMI community to embrace agile. This was in 2004.

I first met Paul at the inaugural SEMAT meeting in Zurich in the spring of 2010. Going into the meeting I knew about three quarters of the people involved, although Paul was one of several people new to me. During one of the breaks we introduced ourselves to each other and started up a conversation. To be honest, my first thought was "Great, another CMMI zealot who wouldn't know what a clue was unless it was documented, reviewed, watered down, reviewed again, and then finally accepted by a committee several months later." But then I listened to him. He had real world experience making CMMI work in practice, often overcoming the well-meaning CMMI true believers whose primary desire was to define and then enforce repeatable processes instead of producing repeatable results. His experiences and opinions were very similar to what Watt's had shared with me, and his observations about CMMI implementations were very similar to my own. More importantly, it was clear to me that Paul had earned his seat at the SEMAT table. Almost immediately after the SEMAT meeting I read Paul's book, *"Integrating CMMI and Agile Development,"* and confirmed that he was a true pragmatist.

Fast forward to the Autumn of 2013. Paul contacts me and asks me to be a technical reviewer of the manuscript of his latest book, this one. In this book Paul provides important insight for succeeding at improving your software engineering processes. Although there is a plethora of great advice in this book, four of Paul's insights struck me as important for achieving lasting process improvement.

First, he rightfully questions the effectiveness of many CMMI implementations. Although this book is about far more than CMMI, I believe that it's critical that we listen to, think about, and then act on the criticisms that Paul shares with us. Paul has worked in the CMMI trenches for years and clearly cares about helping organizations who are on the CMMI path to improve their effectiveness. I have worked with several organizations claiming to be CMMI Level 5 compliant and have been consistently appalled by the wastefulness of their approaches despite their claims to the contrary.

Second, I was struck by his idea of first and second level checks. First level checks are very lean in nature in that they are non-intrusive, support rapid feedback, and support continual small corrections. The goal is to sense, and rapidly correct, commonly repeating weaknesses. Second-level checks enable you to assess whether you are achieving the intended results. These checks should also have a rapid feedback loop to ensure that timely actions are taken. I suspect that you will find that Paul's advice about first and second level checks are easily actionable within your organization.

Third, I found Paul's insights around measurement to be very pragmatic. In particular, his observation that you are often better off analyzing the data that you already have is critical. You have a limited process improvement budget, if you have one at all, so you want to invest it wisely. Investing in more measures, when you likely have a lot of data you aren't yet leveraging, is wasteful. Modern development and operations tools, including open source ones, often generate usage data that can be analyzed and reported on using business intelligence dashboard technology. In fact, this is a strategy called development intelligence in the Disciplined Agile Delivery (DAD) framework.

Fourth, Paul promotes what is effectively a Kaizen-based approach of small changes. This is a strategy that the lean community has promoted for years, Once you have achieved your performance objectives you have to keep making small changes, and you need a mechanism in place to rapidly sense the effects of those small changes, and rapidly respond to those effects to minimize performance impact.

My fear is that many agile transformations in the enterprise space will fail. They will fail because they ignore Paul's advice. When you're transitioning to agile the hardest part is evolving your culture, and unfortunately many enterprises have built a culture for themselves that is almost the exact opposite of agile. This is particularly true in CMMI environments.

Should you read this book? If you are interested in software process improvement, if you are responsible for an agile transformation effort, or if you are an IT professional who wants to get better at what they do, then I also think the answer is a resounding yes. In short, Paul has written another great one.

Scott Ambler, May 2014

Co-Author of *Disciplined Agile Delivery: A Practitioner's Guide to Agile Software Delivery in the Enterprise*

ScottAmbler.com

# Acknowledgements

I thank the following people for reviewing and commenting on the multiple draft versions of this book: Scott Ambler, Dave Cuningham, Bob Epps, Bill Fox, Hillel Glazer, Jim Grebey, Prabhakar R. Karve, Dr. Tom McBride, Jan McMahon, Winifred Menezes, Barry Myburgh, Dr. Pan-Wei Ng, Dr. Michael Oakes, Dr. Cecile Peraire, Dr. Robert F. Palank, Burkhard Perkens-Golumb, Mike Phillips, Jim Steere, John Troy, Dr. Richard Turner, and Dan Williams.

I thank the following colleagues and friends for their stimulating discussions and support for the SEMAT vision: Andrey Bayda, Dr. Arne Berre, Stefan Bylun, Dr. Jorge Diaz-Herrera, Brian Elvesaeter, Tom Gilb, Dr. Michael Goedicke, Dr. Shihong Huang, Ingvar Hybbinette, Dr. Ivar Jacobson, Dr. Bela Joshi, Dr. Mira Kajko-Mattsson, Philippe Kruchten, Svante Lidman, Craig Lucia, Dr Bertrand Meyer, Bruce Macissac, Dr. Jaana Nyford, Gunnar Overgaard, Dr. June Sung Park, Tom Rutt, Ed Seymour, Ed Seidewitz, Dr, Richard Soley, Ian Spence, Michael Strieve, Paul Szymkowiak, and Dr. Carlos Mario Zapata Jarmillo.

Special thanks to my wife Jan for being my sounding board for many of the ideas that went into this book; Scott Ambler for writing the Foreword to the book, and suggesting to include the 15 Fundamentals and Framework Needs in the front of the book; John Troy for his idea to create the framework vision early in the book prior to the discussion of the Essence example; Bill Fox for his idea to write down 100 possible titles for the book and poll the reviewers to find the best candidate; Burkhard Perkens-Golumb for his suggestion to restructure the introduction consolidating my original preface and introduction; Michael Strieve for his help with the Essence pattern diagrams; Bob Epps for his continued life-time of insightful comments and discussions; and Bob Demer my life-long friend who unknowingly inspired me to continue to pursue the personal improvement project discussed in this book.

# Introduction

## The Problem And What This Book Is About

For the past forty years I have been helping high technology organizations in their quest to improve and keep ahead of the competition. During this time I observed millions of dollars spent annually on process improvement initiatives that too frequently fell short of their intended mark. Due at least in part to this situation, today many are turned off and have tuned out when it comes to the multitude of process and performance improvement approaches along with their related hype and buzzwords. Agile [1], CMMI [2], Kanban [3], Lean [4], Six Sigma [5], Lean Six Sigma [6], PSP [7], and TSP [8] to name just a few. We have taken a simple idea and made it far too complex and in so doing have lost the spirit and intent of performance improvement. I have also observed common patterns associated with these past efforts which can shed light on how we can do better at the process of getting better in the future.

## My First Goal in this Book

My first goal in this book is to explain why we are facing these problems and how you can get yourself and your organization back on track focused on the things that matter most to both your own personal performance and your organization's performance. This book is equally about personal and organizational performance.

### *More About the Problem*

Part of the problem we face has been caused by a gap that exists between the theory of performance improvement and the way that theory is being implemented today. As an example, today many organizations that use the CMMI– specifically the high maturity practices– have fallen short of achieving the high value sustainable performance improvements sought [9]. By high value I mean those improvements that address pain points that hurt us the most when we need our performance to be at its best. I am using the CMMI as an example here, but the data presented is reflective of other performance improvement models as well, including Lean Six Sigma [10], and Agile Retrospectives [11], as will be discussed.

Figure Intro-1 demonstrates the improvement theory underlying the CMMI, alongside what **some** organizations have actually experienced [12][1]. I emphasize the word some because there is also growing evidence indicating high maturity practices can reduce cost. [13] Our focus here is on those organizations experiencing difficulties reaping the most valuable benefits from their performance improvement investments, regardless of their improvement approach. I highlight CMMI High Maturity practices because the intent of these practices is to help organizations improve performance and sustain those improvements.

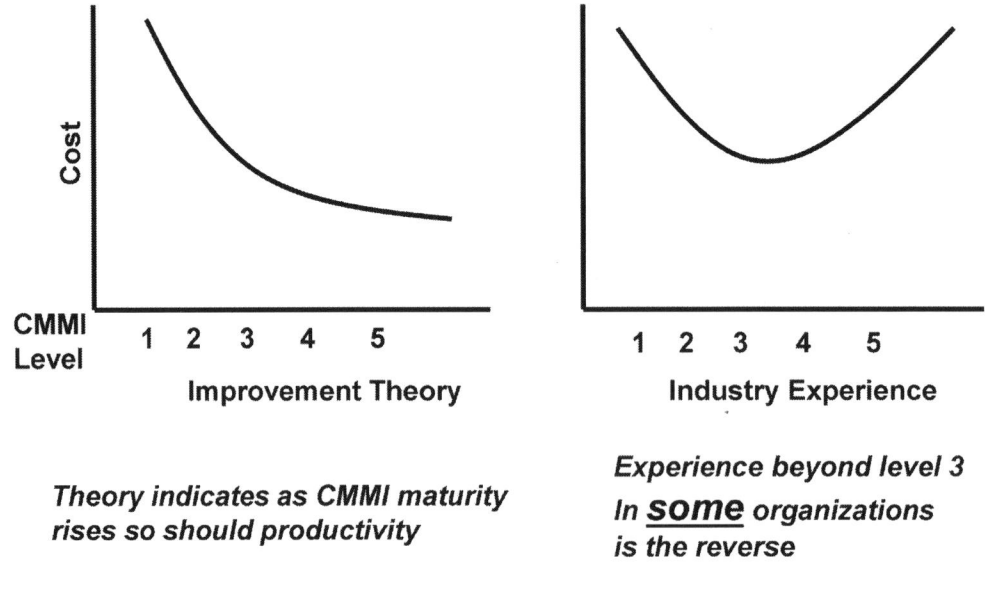

Figure Intro-1 CMMI Theory and Observations

The graph on the left in the referenced figure demonstrates the theory that as CMMI maturity rises cost should decrease (this assumes the same quality/defect level in the

---

[1] Refer to http://www.pemsystems.com/SEMAT_position_McMahon.pdf. Data originally provided by a Senior Manager in a CMMI Level 5 organization.

resultant product). The graph on the right demonstrates that after level 3 cost actually rises in some organizations thus degrading, rather than improving, performance. This data was originally provided by a Senior Manager in a CMMI Level 5 organization and there are numerous other sources available supporting similar experiences [14, 15, 16]. This leads to a question:

*Is the theory underlying the CMMI wrong?*

It is my contention that the theory is not wrong, but as an industry we have not done a good job of translating theory to practice. Or, in other words, we have not done a good job of explaining the theory in simple enough terms so it can be applied easily to everyday situations faced by professionals on the job. Stated another way, the increase in cost observed in the diagram reflects poor CMMI implementation, rather than an error in the CMMI model.

## About the Proposed Solution

To achieve high value sustainable performance improvements in the current technology environment we need a culture shift in how performance improvement is viewed and implemented. Organizations need to take a lesson from how great performers really get better at what they do and apply it within their own business context.

Today many organizations use documented process descriptions, comprehensive formal training, and extensive studies and pilot projects to help their people get better at doing their job. While these techniques can help achieve a level of proficiency, historically they have fallen short at helping us achieve and sustain our most valuable potential improvements taking us beyond fundamentals.

## How Performance (Process) Improvement Works Today

Often a process [2] improvement effort in an organization begins by assigning a small group of employees to the task. The first pattern I have observed is that this group is

---

[2]The focus of this book is performance improvement. Process improvement is a means to that end.

usually secluded from the stakeholders who must eventually use the improvement. This is not wrong. Change must be carefully managed so it does not disrupt critical on-going business. But too often the major portion of these improvement investments is expended by people working the effort in isolation from those intended to receive the help. High value sustainable improvement can only result when those who must perform can utilize the improvement in solving their daily challenges. I have also observed many of these efforts in organizations leading to a cycle that seems to move from one fad to the next with little sustainable improvement[3] over time to show for the effort expended.

If you work in a large corporation, these observations may not surprise you, but have you given much thought to why these patterns occur, and what could be done differently leading to a more sustainable positive result?

In the book "*Talent is Overrated: What Really Separates World-Class Performers from Everybody Else*" [17] the author explores how world class performers move their performance to much higher levels than most people ever achieve through what the author calls "deliberate practice". "Deliberate practice" is not what most of us were taught about practice when we were young. It is highly demanding mentally, and it is not fun.

# A Second Goal in Writing This Book and Why You Should Care

In Chapter One of this book I begin the investigation into what it takes to achieve high value sustainable performance improvement by discussing in more detail the concept of deliberate practice and how it has helped world class performers such as Jerry Rice[4], and Bill Gates. But this book is not about how to become another Jerry Rice or Bill Gates.

On the personal side, this is a book that can show you how to get just a little bit better at whatever you want to get better at regardless of your current performance

---

[3] When I use the phrase "sustainable improvement" I mean improvements that last without reverting to old behaviors.
[4] Jerry Rice is a retired American Football wide receiver who is generally considered to be one of the greatest players in National Football League history.

level, and it will show you how to sustain that performance improvement once you have achieved it.

On the business side, organizational process and performance improvements can be broadly grouped into two categories: Technology and major organizational changes looking to the future, and smaller changes that can help practitioners every day.

A second goal in this book is to present the case that the speed of change we are all witnessing in today's world requires a rebalancing of how organizations view and prioritize their process and performance improvement initiatives across these two broad categories.

This may not sound like a very big goal for a book, but think about what it would mean to you and your organization if everyone just got a little better at what they did today, and tomorrow everyone got a little bit better again.

This book provides practical techniques that have been proven to help individuals and organizations get better and sustain their performance improvements into the future.

# The Approach in this Book

Before we jump into the material, let me tell you a little about the approach I took in developing the book. This book is not about any specific performance improvement approach, but it does discuss many popular approaches including the CMMI, Lean Six Sigma, and Agile Retrospectives. My approach is to highlight fifteen (15) fundamentals I have observed common to all successful improvement efforts where sustainable high value performance improvements are achieved. These fundamentals too often get missed by organizations and individuals trying to improve and sustain their improvements.

The intent is not to focus on, or dive too deep into, any specific improvement approach, but to highlight what is common across all of them with respect to the essentials of effective performance improvement.

In this book I share real examples from both my own consulting experiences, a personal performance improvement experience, and stories from high performing athletes and musicians to help you think about performance improvement outside-the-box.

## A Third Goal of this Book

A third goal of this book is to share a vision for a framework that can help counter the patterns that may be holding you and your organization back. I am not talking about yet another buzzword, method, or new tool to hype, but rather a simple thinking framework that can help keep you and your organization focused on the fundamentals too often missed.

## Summarizing the 15 Fundamentals and Thinking Framework Needs

Below you will find a summary of the 15 fundamentals and the related thinking framework needs in support of each fundamental. In order not to distract the reader from the main flow of the book I have chosen to make observations about this framework in sidebars (noted by "Framework Vision") in Parts I and II of the book. In these sidebars I explain the framework's key characteristics, rationale for the characteristics, and connection to the fundamentals.

In Part III of the book I share an example of such a framework that holds promise and I explain how this framework may be able to help organizations achieve and sustain the higher performance they seek.

### Fundamental One:

Training is about helping people understand expectations related to a job. Practice helps you actually do your job, and learn to repeat how you do it, and continue to do it well even under difficult and often unanticipated conditions.

Related Thinking Framework Need: Our framework envisions practices as living entities reflecting what people actually do. With our vision your practices are built as extensions to a set of essentials that have been widely agreed upon.

Learning to perform a practice effectively requires more than just acquiring knowledge about our processes. It also requires an understanding of the context we must perform within and it requires that we learn how to make rapid and effective decisions that fit within a specific context. Thus, our framework must support these real world needs.

## Fundamental Two:

We each have tendencies toward repeating specific weaknesses, and experience has shown they are often the largest obstacle people and organizations face when trying to sustain higher levels of performance. A critical first step toward sustaining higher performance is to locate areas that contain critical repeating specific weaknesses that hinder your personal or your organization's performance.

<u>Related Thinking Framework Need</u>: A key difference with our framework vision from what has been done in the past can be summed up in the phrase *"separation of concerns"*.

Our goal is to separate the essentials (e.g. what all successful projects should focus on) from extensions (e.g. specific practices, such as those needed to address a specific situation or weakness).

The rationale for this goal is to simplify the management of practices for individuals, and teams allowing them to keep their specific practices up to date reflecting their specific needs based on their specific situation, without having to worry about whether their changes are jeopardizing the essentials, that we all need to be constantly focusing on.

## Fundamental Three:

One of the best ways to keep people motivated and interested in their work throughout their career is to involve them in their own continuous improvement.

<u>Related Thinking Framework Need</u>: Our framework vision places the professional in charge of their own practices, supports them in identify where changes are needed, and empowers them to make those needed changes. Some fear this approach believing that project pressures may lead personnel to make poor decisions degrading rather

than improving their practices. This is one of the reasons why separation of concerns is so important. Separating the essentials from specific practice extensions ensures the essentials are never lost as we make changes to improve and sustain our higher performance.

## Fundamental Four:

You have to figure out your own repeating specific weaknesses if you are to gain the benefits and achieve your own sustainable higher performance.

Related Thinking Framework Need: While each organization must locate their own unique repeating trouble spots, our vision includes a framework that can help organizations by guiding them to consider areas where trouble has most often been found in the past.

## Fundamental Five:

You should always, as individuals, reduce the number of areas you are focusing on at any one time to between three and seven, ideally closer to three. Organizations can have more, but this should not increase the focus of individual performers.

Related Thinking Framework Need: Our vision stresses that all changes to your way of working should be accomplished incrementally in small steps supporting fundamental five.

## Fundamental Six:

When selecting areas to measure ensure you have done sufficient analysis (e.g. following real threads) to know you understand the real context and there is high likelihood of findings that will lead to improved performance in the reasonably near future.

Related Thinking Framework Need: Our framework needs to be a "thinking framework" in the sense that it helps individuals and teams make better decisions related to small timely changes by providing the team with objective data they can use to support their decisions.

## Fundamental Seven:
Some of the most significant impacts to performance start out as seemingly little things that we often fail to notice until it becomes too late to correct.

<u>Related Thinking Framework Need</u>: Helping practitioners make tough choices where they don't have enough time to do everything may be the most valuable area where our thinking framework can benefit our teams.

## Fundamental Eight:
When you've been doing something wrong for an extended period, the right way may feel wrong for a period of time while you are adjusting to the change.

<u>Related Thinking Framework Need</u>: We envision the framework helping most at the start of a project (e.g. preparation steps). If the project runs smoothly, its value may diminish as the project proceeds. On the other hand, if the project starts to run into trouble, the value of the framework rapidly rises.

This is because the framework is a monitoring aid. You can liken it to a good referee in a sporting event. In well played games good referees are often not noticed. Their value becomes clear when the trouble starts.

## Fundamental Nine:
Collecting more and more samples of the same data won't help an organization improve or sustain higher performance.

<u>Related Thinking Framework Need</u>: The framework will remind the team where they are and where they need to focus their effort next.

### Fundamental Ten:

Often the best path to high value performance improvements, especially when you have a limited process improvement budget, is to spend less time collecting data, and more time analyzing what data you have collected, and then using the results of that analysis to keep refining your resolution and measurements to ensure you are moving in the right direction.

Related Thinking Framework Need: You may decide there are other important things you work with that you want to monitor and progress, beyond the essentials defined within the framework.

Therefore the framework will allow you to add your own important things to monitor and progress.

### Fundamental Eleven:

Just following a process isn't enough to sustain high performance. It must be the right process that addresses the real goal.

Related Thinking Framework Need: The framework will not magically give you answers to all your challenges, but it will provide a practical and simple way to rapidly remind people what is most important to be focusing on right now, and it will provide a structure under which you can add more specific information to help you find your own answers to your challenges.

The rationale for this framework need is based on the observation that too often practitioners are faced with too much work on their plate and they often need help in deciding where the priority should be placed.

### Fundamental Twelve:

Even if we know it makes sense to practice, it won't help if we don't discipline ourselves to do it consistently at the right time.

Related Thinking Framework Need: The framework must be easy to access, use and update as practitioners learn new things interacting with their teammates each day on the job.

The rationale for this framework need is based on the observation that if it isn't easy to access, use and update, it simply won't be used by busy practitioners.

## Fundamental Thirteen:

Once you have achieved your performance objectives you have to keep making small changes, and you need a mechanism in place to rapidly sense the effects of those small changes, and rapidly respond to those effects to minimize performance impact.

Related Thinking Framework Need: The framework supports fundamental thirteen by placing practitioners in control of their own practices and giving them a mechanism to keep their practices current with the information they need to continue to maintain high performance.

## Fundamental Fourteen:

The most valuable performance improvements often involve situations that seem to defy resolution because there is no quick fix, but we know they are critical to our performance and we know there is no way to work around them so we must continually deal with them head on.

Related Thinking Framework Need: The framework will provide reminders to practitioners of common situations they should be alert to, and possible options and consequences to potential related decisions.

## Fundamental Fifteen:

Just knowing what is happening isn't enough to keep it from happening again. You must practice continually at just the right time with the right objective and contextual data, if you really want to make the changes necessary to sustain higher performance.

Related Thinking Framework Need: The framework will focus on the most important things we work with. It will help our teams assess their progress in a consistent agreed to way reminding them when they need to rapidly respond with a change.

# Neuroscience, Human Decision-Making and the Stories in this Book

Today through recent breakthroughs in neuroscience we understand more clearly how human decision-making occurs [19, 20], and through the stories in this book you will learn how to leverage this new information about decision-making to help the performance of your people and your organization.

# This Book is for You if…

You are an organizational leader, process improvement professional or software or systems practitioner and…

- you believe there is no single best approach to performance improvement.[5]
- you believe the best approach to performance improvement includes conscious thought and clear decision-making integrated into every team and practitioner's way of working.
- you want to learn why many performance improvement efforts fall short of their goals so you can avoid similar pitfalls.
- you are interested in learning fundamentals that are common to all successful improvement approaches, but are often missed.
- you are ready to do some out-of-the-box thinking related to the performance improvement problem we all face today.
- you are interested in improving both your own personal and your organization's performance.
- you are interested in seeing practical and easy to understand examples of high maturity thinking (although some may be non-traditional) that can bridge traditional and agile approaches.
- you are interested in learning about proven techniques that can help both your organization's performance and your practitioner's performance.
- you believe the best path to high value sustainable performance improvements is incremental, continuous and must involve practitioners in a more active way than what most organizations have done in the past.

---

[5]For a cross-reference of CMMI, Lean Six Sigma, and Agile, specific topics in this book refer to the Appendices.

# About the Terms Process and Practice in This Book

The CMMI defines the term "process" as "a set of interrelated activities which transform inputs into outputs to achieve a given purpose. " The term process has historically been associated with the written description of the process. Some organizations use the term practices the same way. That is they view their "practices" as a description of how they would like their people to behave. But the practices (or processes) that an organization actually follows may not be written down. They may be tacitly known and followed by the people in the organization. In this book when I use the term "practice" or "process" as a noun I mean it to include both types–written and tacit.

When I use the term practice as a verb, I mean the activity of rehearsing your practices. Do you think you need to rehearse your practices to gain the potential high value payback? Or do you think you just need to be given a little training in your practices before you follow them for real and then you will naturally get better through experience alone?

If you believe that preparing is important before you perform, then how should you go about doing it, when should you do it, and how much of it should you do, if you want to maximize a high value payback for your effort? These are questions that rest at the core of the subject matter in this book.

# Part 1 – The Problem & The Start of a Solution

# Chapter One – The Sustainable Performance Dilemma

*"Out of intense complexities, intense simplicities emerge." Winston Churchill*

When I was young my goal was to become the world's greatest golfer. In the summer time my father would take me very early every morning to a small nine hole golf course in Windsor, New York. I would stand out on the far side of Route 79 from where the first hole was situated and hit golf balls to the other side of the road. It was not uncommon for me to hit 200 golf balls before nine in the morning. Then I would practice putting and chipping for an hour before heading out to play nine holes. In the afternoon I would repeat the entire cycle.

I had been taught if you want to get better you had to practice long and hard. There was no big secret to it. The harder you practiced the better your chance for success. That was what I was taught, and that was what I believed then. Today I no longer believe it, and if I knew then what I know now about high value performance improvements I believe I would have had a much better chance of achieving my goal.

## What Is Practice?

We all know what practice is, or at least we think we do. It's what athletes and musicians do. Right? In the book *"Talent is Overrated"*[17] the author, Geoff Colvin, describes what practice used to mean to him as follows: *"When I practice golf, I go to the driving range and get two big buckets of balls. I pick my spot, put down my bag of clubs, and tip over one of the buckets. I read somewhere that you should warm up with short irons, so I take out an 8 or 9 iron and start hitting. I also read somewhere that you should always have a target, so I pick one of the "fake" greens out on the range and aim for it, though I am not really sure how far away it is. As I work through the short irons, middle irons, long irons, and driver, I hit quite a few bad shots. My usual reaction is to hit another ball as quickly as possible in hopes that it will be a decent shot, and then I can forget about the bad one. Occasionally I realize that I should stop to think about why the shot was bad. There seems to be about five thousand things you can do wrong when hitting a golf ball, so I pick one of them and work on it a bit, convincing myself that I can sense improvement, until I hit another bad one, at which point I figure I should probably also work on another one of the five thousand things. Not long thereafter the two buckets of balls are gone and I head back to the clubhouse, very much looking forward to playing an actual game of golf, and feeling virtuous for having practiced. But in truth I have no justification for feeling virtuous. Whatever it was I was doing out on the range, and regardless of whether I call it practice, it hasn't accomplished a thing."*[6]

> ## Framework Vision: View of practice
>
> Historically, many organizations have viewed their practices as static descriptions of the way they would like their people to operate. Over time, because people constantly learn new and better ways to do their job, a growing gap has resulted between these "practice descriptions", or "shelf-ware" and what people actually do on the job. Our framework vision takes a different view of practices. We envision practices as living entities reflecting what people actually do rather than what someone thinks people should do. With this vision your practices are built as extensions to a base of knowledge that has been widely agreed upon. While this vision may sound ideal, we also understand that in the real world, learning to perform a practice effectively requires more than just acquiring knowledge about our practices (or processes). It also requires an understanding of the context we must perform within and it requires that we learn how to make rapid and effective decisions that fit within a specific context. Thus, our framework must support these real world needs as well.

Does this sound like what you think practice is? Colvin discusses how great performers such as Jerry Rice, and Bill Gates achieved greatness by practicing in a different way that he calls "deliberate" practice. Deliberate practice has the following characteristics:

---

[6] "Talent Is Overrated," by Geoff Colvin, page 65

- It is designed specifically to improve performance
- It can be repeated a lot
- Feedback is continuously available
- It is highly demanding mentally
- It isn't much fun

When I read those characteristics I wasn't surprised. They sound more like what I was brought up to believe about practice than what Colvin had described his view of practice had been. I was taught you had to dedicate yourself to your practice. You had to totally commit yourself to your practice everyday giving it everything you had from sunrise to sunset. And you needed to keep doing it even when your body wanted to stop. That is what I believed and that is what I did, but it didn't work for me. It wasn't enough to sustain my continued performance improvement. Besides failing to achieve my performance goals, when I was about twenty years old I stopped playing golf altogether and I didn't play the game again for almost thirty years.

# Looking Back At Yesterday And At Business Today

I believe there were many factors involved in why my hard work didn't lead to achieving my golf goal. When I would spend long days practicing I frequently became tired in the late afternoon and found it difficult to keep my mind on what I was doing. But since I was taught that practice isn't suppose to always be fun I would keep hitting balls. Often as a result my golf swing would actually degrade and by the end of the day I would be hitting the ball worse than when I started early that morning. I felt like I was on a yo-yo, seeing improvement for a period, followed by loss of improvement. While progress was clearly evident in my early years, over time I reached a point where it seemed that the more I practiced the less value I was getting out of it and the more frustrated I became.

Today in business I see similarities to this cycle where organizations improve to a certain level and then hit a downward period even though they might still be working hard at trying to get better [9]. There are usually multiple factors involved as to why this pattern occurs, but my observation in many organizations is that sustainable improvement beyond a fundamental capability is not implemented effectively and therefore is rarely achieved. I find this to be true even in CMMI Level 5 organizations which are supposedly continuously improving their processes and hopefully demonstrating valuable performance improvements as a result [21]. While it is natural to expect a degree of up and down cycles, I believe the impact of the downward cycle in many organizations is far greater than what many may realize.

# You Can't Sustain Performance Staying Where You Are

One of the most common misunderstandings about performance relates to what it takes to maintain a specific performance level. You can't sustain your current performance by doing nothing about improvement. This is not well understood. Without working improvements you will lose ground. This is because the environment in which you operate never stays the same. Personnel may come and go. Projects end, others start. With the passage of time, if nothing else changes people forget things they use to know.

The specific factors that cause loss of capability will differ in each organization, but loss always occurs and it is almost always more costly than many realize. An observation I made both with organizations and individuals with respect to loss of capability is that over time there are performance patterns that tend to repeat. Once you understand these repeating patterns for a given organization, or individual, you can start to predict what lies ahead. More importantly to our purpose here, you can also learn how to change the pattern so your losses are mitigated and offset by improvements that allow you to continually sustain higher performance in the future.

Most people understand that process improvement and training are important. But many don't understand how effective sustainable performance is developed and maintained in an organization. Where this becomes most evident is in the decisions made when the business is in one of its normal down cycles showing signs of trouble. For example, when sales are slipping, or when projects are behind schedule.

It is often under these conditions where the benefits of your most recent process improvement efforts could potentially be realized and be of most help to you. But this is what is most often missed. This is because many don't understand how high value sustainable performance improvements are attained and nurtured.

If you look at the athletes who most often achieve success you will find they are the ones who are best prepared because they trained in conditions similar to (or more difficult than) those in which they must perform during competition. The best athletes seek out the most difficult conditions in which to practice and improve their performance. Runners who practice in cool weather wither on race day when the temperature turns hot. West Coast professional baseball teams rarely perform well when they head East for the World Series in cold October. Successful performers seek out adverse conditions in which to practice because they know these conditions provide the optimum environment to prepare to perform at their best.

But unfortunately rather than greet the difficult conditions when they arrive in business as an opportunity to help personnel learn to perform at their best, we find more often these are the conditions under which investments in people most often dry up. There are, of course, reasons why this happens. For example, an organization's motivation to improve performance may change over time because of changing business conditions. But should changing business conditions cause us to lose our motivation to improve, or just refocus us on the high value areas given the current environment? This will be discussed further, along with what you can do to sustain your motivation, as we move forward in the book.

## Patterns During Difficult Times

I observe repeating patterns in each organization I am asked to help. Faced with a crisis, in some organizations project management becomes engaged driving a solution. One pattern I have commonly seen, however, is for management to become overly engaged in the solution resulting in a lack of leadership and guidance deep in the organization during the crisis.

In other organizations engineering takes the reigns in times of trouble driving a solution from the technical side. A common pattern I have learned to be on the look out for in strong engineering organizations has been a loss of cost and schedule accountability during these times of high technical focus.

During times of crisis organizations that have recently instituted improvements to solve common weaknesses often revert to old habits leading to loss of performance—rather than looking to the crisis as an opportunity to apply the new behavior to gain high value performance benefits.

The manner in which process improvement is viewed and managed in most organizations today supports this backward mentality. This is because of the way process improvements are most often implemented today as distinct and segregated efforts from real projects. This makes them easy targets for elimination in these situations where they could potentially provide their greatest benefit. But when such decisions are made do we understand the consequences not just to potential future improvements, but to those investments previously made and to our current on-going critical projects?

Observing these common patterns has led me to ask questions related to what effective sustainable performance requires, and this investigation has led back to the notion of practice–but not the kind of practice I learned in my youth, nor the kind attributed to the success of our superstars, like Gates or Rice.

# The View of Practice in Many Organizations Today

When we think of someone needing to practice a subconscious thought many of us have is the need to practice implies a lack of competency. For athletes or musicians practice does not bring this negative connotation. We accept that getting better for them is a noble goal. We understand you can always get better as a musician, or an athlete and if you don't practice we know you won't perform as well as you could during the next game or performance.

But why do we take a different view when it comes to people in the business world? This same idea of needing to practice to ensure you are ready to do your best on the next project, or the next phase of the current project, or even the next day of the current project doesn't usually sit so well with the way we think. In fact, in business the idea of practice takes on more of a negative connotation. Rather than look upon practice as something that improves performance in the business world we tend to think that if someone needs to practice then:

*"I don't want them doing it on my project because I want someone who already knows how to do the job".*

The thinking for some reason in business becomes either you are qualified to do the job, or you are unqualified. We don't tend to think in terms of always being able to perform at one's best by practicing in the context of work. But why should there be any difference?

In fact in business practicing on the job is likely to be discouraged because it may mean if you are trying to do it better tomorrow you are risking doing it right today. We don't want to take that risk. This has historically been one of the prime reasons why deploying process improvements in the business world is often resisted. While our top athletes and musicians are constantly working to get better even right up to a few moments before a big game or performance, in business we avoid change fearing loss of performance rather than recognizing the great potential opportunity it could bring.

One of the arguments I often hear is that we send people to training so they shouldn't need to practice on the job. But this line of reasoning assumes that the purpose of training and practice are one and the same. This misses a fundamental value of practice that training does not provide.

## Fundamental One:

Training is about helping people understand expectations related to a job. Practice helps you actually do your job, and learn to repeat how you do it, and continue to do it well even under difficult and often unanticipated conditions.

Practice under adverse conditions is required to help us learn to do a job in the environment we will actually have to face when performing that job. Today in business because practicing "on-the-job" is discouraged we often don't get to practice at all for the real world conditions we are asked to operate in. As a result—like the runner who practices in cool weather and then must face sweltering heat on race day– we are not prepared and time and time again fall short in sustaining our most valuable potential performance improvements when they could help us the most.

## Starting To Discover A Different Kind of Practice

When I first began to think about a better way to conduct performance improvement efforts, and if it might tie to practice, I was working with a successful growing organization helping them deploy formal training of some recently improved processes. While they liked the formal training there always seemed to be people who missed the scheduled training because it was only offered at certain times. We also received feedback that while the training was good, there were frequently issues on projects that people didn't know how to handle. These were often very *specific issues* such as when and how a certain type of risk should be raised to senior management, or how to handle a specific subcontractor issue, or a specific customer issue.

Often the answer to these specific issues being raised was actually in the training material, but it required an interpretation on *how to* apply a fundamental practice [22] [7] we had previously taught to a very specific situation. This was something we found many people required additional help doing.

### Framework Vision: What's different?

Helping people handle the different real situations that arise on the job is itself a process improvement. It is also one of the best ways to aid real on the job performance. Unfortunately, it isn't viewed this way in many organizations. Once the fundamental process is defined with basic training, organizational investment in that process often dries up. What makes this so unfortunate is that the investment stops just at the point where the potential performance improvement starts. This is why in organization's that use the CMMI framework stopping at CMMI Level 3 (basic processes defined) is not a wise business investment. Level 3 means you have a base from which you can start serious performance improvement efforts. What is different in our vision is that we do not try to separate process (or practice) definition from practice improvement and practice evolution. Improvement is continuous and integral to practice execution. This is a fundamental difference that is necessary to achieve real sustainable performance.

---

[7] Disciplined Agile Delivery (DAD) talks about addressing process goals, not following practices. The idea is that your team will do so in a way that reflects the situation they face. Refer to http://disciplinedagiledelivery.wordpress.com/2013/07/17/exploring-initial-scope-on-disciplined-agile-teams/ and http://disciplinedagiledelivery.wordpress.com/2013/01/21/disciplined-agilists-take-a-goal-driven-approach/

# Sustainment Training

Because these situations were arising frequently one of the managers suggested that we consider holding shorter, less formal, and more frequent sessions that came to be called *Sustainment Training* sessions. The idea of sustainment training wasn't that it provided new training, but it was refresher training for just key issues that were reoccurring where people needed *reminders*, and could get *specific questions answered*. They needed help, and in some cases they needed specific guidance on scenarios that were specific to their projects. These sustainment training sessions I came to view more like *"practice"* than formal training.

We were repeating principles and practices that we had already taught, but we were going **much deeper into "how-to"** apply the principles and practices by looking at actual *scenarios* that were occurring on real projects and then considering different options and consequences to handle each.

These sustainment training sessions proved to be extremely popular in helping people deal with the actual situations that were arising on their project. The sustainment training we implemented had some similarities to Colvin's *"deliberate practice"* and some distinct differences. Sustainment training was similar to *"deliberate practice"* in that:

- It was designed specifically to improve performance by taking into consideration the real situations people were facing everyday specific to their projects
- It can be repeated a lot and we found it needed to be repeated a lot because without repeating it, it became too easy for people to forget or just not to see the connection between the practice scenario and their real project situation
- Feedback was continuously available because we started holding these sessions more and more frequently opening them up for anyone to just stop by during their lunch break to get some quick feedback, and reinforcement

Sustainment practice was also different from *"deliberate practice"* in that:

- It didn't take long and therefore wasn't demanding from a time perspective
- It was fun, because we made it fun. It became a period for sharing across groups which improved morale in the company.

More and more people just started bringing their lunch and attending to listen even if they didn't have questions. They looked forward to these mid-day sharing experiences.

When I think back to my youth, I thought I was practicing the right way to get better at golf. But my practice wasn't effective and it didn't feel right. I knew there must be a better way to get better, but I didn't know what it was, and I didn't know how to go about finding it.

# Chapter One Summary Key Points

- You can't sustain your current performance by doing nothing about improvement.
- Training is about helping people understand how to do a job.
- Practice helps you apply that training, and continue to apply it even under stressful conditions.
- Helping people handle the different real situations that arise on the job is itself a process improvement. It is also one of the best ways to provide high value assistance to performers on the job.
- CMMI Level 3 means you have a base from which you can start serious performance improvement efforts. Too few organizations actually understand and do this.
- Deliberate practice characteristics
    - It is designed specifically to improve performance
    - It can be repeated a lot
    - Feedback is continuously available
    - It is highly demanding mentally
    - It isn't much fun
- Sustainment training similarities to deliberate practice
    - Designed specifically to improve performance
    - It can be repeated a lot
    - Feedback is continuously available
- Sustainment training differences from deliberate practice
    - It doesn't take long
    - It is fun

# Chapter Two – Repeating Specific Weaknesses/ Why You Should Care

*"What gets measured gets managed."* Peter Drucker

In this chapter we begin an investigation into an idea I recently discovered not through any books, but by personal experience. This idea I believe is essential to sustainable high performance, although it isn't well known or given much attention when applying most of today's popular performance improvement approaches. I call it *repeating specific weaknesses*. You can think of repeating specific weaknesses as a special category of "*pain points*" that need to be identified and continually managed if you are serious about sustaining higher performance.

You may be thinking right now, "*I am serious about sustaining higher performance, but my manager keeps changing my goals.*"

As we start this chapter I explain what led to the idea of repeating specific weaknesses, and I explain why individuals and organizations serious about high value sustainable performance improvements– especially in an environment where the goal keeps changing– need to start by first identifying their own unique repeating specific weaknesses. This is a crucial first step because to be successful in your performance when the environment around you keeps changing you need to constantly be sensitive to staying focused on the right things.

In this chapter I also introduce two performance improvement projects– one personal, one business– that we will return to multiple times in subsequent chapters to help you understand why traditional approaches to handle common trouble spots fail with repeating specific weaknesses. Lets start with a personal improvement project.

## Background for Personal Improvement Project

When I was about 50 years old I started playing golf again when my son asked me to teach him how to play. Ten years later in 2008 a friend asked me to go on a full week golf trip to celebrate his 60th birthday. I agreed, but did so reluctantly because to enjoy playing that much golf I felt I had to perform to a level of proficiency I didn't think I could attain given the time I had available. So the challenge I faced was to get my game where it needed to be in a short period of time. To address this challenge I decided to use a similar approach that I use when helping clients who have similar improvement challenges and resource constraints.

## Business Approach to Performance Improvement

In my business when I am asked to help an organization improve I always start with what is referred to as a "gap analysis" [23]. A gap analysis identifies the "gap" between where the organization currently is and where we want "to be." To determine the current "as-is" state I conduct interviews talking to people to find out how they do their job. In these interviews I ask open ended questions allowing people to just talk, and as they do I listen and take plenty of notes.

The reason I do this is because, while most organizations have plenty of weaknesses, I have found that when you let people talk about their job you hear key patterns that lead to a small set of repeating specific weaknesses that identify the areas where we can gain the greatest performance boost for the limited time and effort available. This is because these weaknesses keep coming back and they tend to repeat most often during critical times when they can hurt your performance the most.

I have also found these weaknesses are unique to each organization and more importantly by focusing on these special pain points we have our best chance to help the organization achieve its performance improvement goals. A simple example of a common repeating specific weakness could be an organization that has a habit of releasing a software product to the field with inadequate testing due in part to pressures exerted from the sales department and/or poor testing practices.

# Comparing Repeating Specific Weaknesses to Normal Process Variation

All processes have some variability. In other words they have certain attributes that can take on an acceptable range of values. For example, when two people are given similar tasks, due to differences in experience, skill levels, and problem solving styles, there will be differences in how they approach their task leading to a range of acceptable results.

Allowing a degree of freedom in how people solve problems is good as it supports accessing the personal strengths of individuals. However, when the variation leads to unacceptable results, some form of intervention is often needed, such as training or mentoring, to bring the variation back within acceptable limits.

Repeating specific weaknesses are different from these normal process variations in that they present serious obstacles to each individual, or organization, holding them back at critical times from achieving their goals. They are also different from normal process variation in that standard training and mentoring approaches have proven insufficient at overcoming these obstacles.

# More About the Personal Improvement Project

On my personal improvement project in 2008 I first conducted a "gap analysis" to determine the "as-is" state of my golf game by taking notes related to how I felt about my game after playing a few rounds. I then analyzed the notes which led me to recall three common problems I used to have when I was younger that my golf teacher always had to remind me of. All three problems related to my preparations (or set up) to hit the golf ball. As it turned out, by focusing on just those three weaknesses and figuring out how to sense them and keep them under control, I was able to improve my performance in just a few weeks beyond what I had been able to achieve in the ten previous years by using the conventional approach of focusing on fundamentals to get better.

> **Glancing Forward: Sustaining performance**
>
> You will learn later in the book that the experience described here worked to help sustain the level of golf I desired for the full week with my friend. However I was unable to sustain that same level of play during the following months and years. By maintaining records of my golf activities, subsequent analysis demonstrated a pattern that led me to discover three years later what I needed to do to sustain my performance over a much longer period of time. This crucial piece of information will be explained along with techniques that can help you apply a similar approach to sustain your own performance.

My personal and consulting experiences have led to the next fundamental to achieving and sustaining higher performance.

## Fundamental Two:

We each have tendencies toward repeating specific weaknesses, and experience has shown they are often the largest obstacle people and organizations face when trying to sustain higher levels of performance. A critical first step toward sustaining higher performance is to locate areas that contain critical repeating specific weaknesses that hinder your personal or your organization's performance.

> **Sidebar: Lean techniques, bottlenecks, and repeating specific weaknesses**
>
> Repeating specific weaknesses can be thought of as a special category of trouble spots or common bottlenecks. We can broadly group common trouble spots into two categories; those that can be solved with a single action, and those that can't because there isn't a single quick fix. With lean techniques once you remove a bottleneck you move on to the next bottleneck. Unfortunately due to the nature of repeating specific weaknesses this simple approach doesn't work because the problem keeps coming back.

# A Business Example of a Repeating Specific Weakness

Let me share a business example of a repeating specific weakness in order to provide insight into the performance sustainment challenge often faced on projects. One of my clients was experiencing difficulties consistently meeting their customer cost and schedule commitments. I was asked to help by conducting a root cause analysis. So I started by talking to people in the organization and gathering some facts. This investigation led to the procurement department and the process to get hardware ordered and installed. By following a few detailed threads we were able to isolate a number of cases of late hardware orders as caused by missing data on a procurement requisition and an inexperienced procurement specialist who didn't know what actions to take when certain data was found missing.

In this case we provided some direct coaching to the new employee so he would better understand his options if the same situation occurred in the future. However, as it turned out, it wasn't always the same data that was missing on the requisition form, and the right action to take when different pieces of information were missing turned out to be different. So while we had identified a specific area that was causing this organization pain, the exact resolution could not as easily be identified and put in place with a single quick fix.

> **Sidebar: Two Types of repeating specific weaknesses**
>
> Repeating specific weaknesses can be broken down into two types. The ones that can't be solved because they are actually more that one problem (later in the book I refer to this as Type 1 repeating specific weaknesses). The business example presented in this section falls into this category. The other type (referred to as Type 2 later in the book) includes situations that repeat due to changing project conditions which tend to mask the problem.

We will return to this example in later chapters when we discuss in more depth how one solves repeating specific weaknesses and their relationship to Causal Analysis and Resolution (CAR), a CMMI Level 5 Process Area. I now want to motivate why repeating specific weaknesses are critical to sustainable performance.

# Repeating Specific Weaknesses: Critical to Sustainable Performance

Let me start by raising two questions:

*Why is it important to identify repeating specific weaknesses?*

And:

*Shouldn't we really be focusing on our strengths, rather than our weaknesses?*

The idea of identifying repeating specific weaknesses is not something I learned when I was young. I wasn't even aware when I was young and playing golf that I had these weaknesses. Nor did I learn about this idea in school or when reading any books on how to get better at golf or any other activity. In fact much of what I had learned when I was young led me to ignore my weaknesses and focus on fundamentals that the experts all agreed upon.

"*Focus on the fundamentals. Keep practicing so you can 'groove' your swing. Keep working hard.*"

This was the mantra I kept hearing and believed in as the best route to achieve my goal.

When I began helping high tech organizations with their own performance goals most of the books I read focused more on strengths than weaknesses. In the book, *"First, Break All the Rules"* [24] the authors who based their findings on in-depth interviews by the Gallop organization of over 80,000 managers in 400 Companies indicate that the best managers don't worry about an individual's weaknesses, but rather focus on their strengths. I actually agree with the ideas in this book. So why am I suggesting here that we should be focusing on our weaknesses?

While I agree with the authors of *"First, Break All The Rules,"* and I often make similar recommendations related to focusing on the strengths of people, I have also learned through experience that there is an appropriate time to focus on strengths and an appropriate time to focus on weaknesses if you really want to attain sustainable higher levels of performance.

I have also found that there exist certain weaknesses that can be worked around and therefore require less attention. However, as we will see when we examine more closely the characteristics of repeating specific weaknesses, they have a critical quality of hurting our performance at just the wrong time, and so if we don't do something about them we will find ourselves– as I found myself when I was young and practicing my golf– on that frustrating yo-yo getting better for a period of time, followed by degradation of performance. Now lets talk about the relationship between fundamentals and repeating specific weaknesses.

# Fundamentals First, But Why They Aren't Enough

It is important to understand that I am not discounting the value of fundamentals and focusing on strengths. Both are essential to sustainable high performance. Fundamentals are important in any endeavor and you should learn the basics of what you are doing first. This is true for golf, tennis, other athletic activities, musical instruments, and business skills. In business we hire people who have gone to school to attain necessary fundamental skills (i.e. accounting, finance, engineering, management) and then you may take more specific training courses within your organization to help you understand the specific processes you must conduct within your organization. But all this will only get you so far. Our careers in the workforce are likely to last 40 years or more. This is historically much longer than the time we spend acquiring our formal education. So after you have taken your college courses, and the initial internal training in your Company, how do you keep getting better at what you do?

> **Framework Vision: Separating essentials**
>
> A key difference with our vision from what has been done in the past can be summed up in the phrase *"separation of concerns"*. Our goal is to separate the essentials (e.g. what all successful projects should focus on) from extensions (e.g. specific practices, such as those needed to address a specific situation or weakness). The rationale for this goal is to simplify the management of practices for individuals, and teams allowing them to keep their specific practices up to date reflecting their specific needs based on their specific situation, without having to worry about whether their changes are jeopardizing the essentials that we all need to be constantly doing.

This question begs us to ask a more fundamental question:

*Is it reasonable to think that people should keep getting better at what they do throughout their career?*

I believe most would agree the answer to this question is yes. We all should be life-long learners.[8] But that brings us back to just how one goes about continually getting better at what they do. Today I would argue there is not an effective strategy in most organizations to help their employees continue to get better at what they do throughout the full extent of their careers. Many organizations do have on-going training. This may include refresher training, or training in new approaches to keep people current with the latest techniques and available tools. But most often this training is general in nature and doesn't address the specifics of an individual's job. I often hear from employees inside client organizations about the frustration they feel

---

[8]This is not intended to imply necessarily a life-long focus on the same career. People may make dramatic shifts throughout their career, often as the result of what they have learned previously.

due to the lack of meaningful help in the *how-to* or *how-much* specifics related to their job.

## Keeping People Motivated Throughout Their Careers

The reason this topic should be of great concern to both individuals and organizations is because one of the greatest challenges organizations face today is how to keep good people from leaving the organization. How do you keep them motivated and interested in their work throughout the full life of their careers? This brings us to the next fundamental.

### Fundamental Three:
One of the best ways to keep people motivated and interested in their work throughout their career is to involve them in their own continuous improvement.

I have found repeating specific weaknesses occur both in personal activities such as playing a sport, or a musical instrument, and I have observed them inside most of my client organizations. I have also found that repeating specific weaknesses are *unique* to each individual and each organization. This brings us to another fundamental.

### Fundamental Four:
You have to figure out your own repeating specific weaknesses if you are to gain the benefits and achieve your own sustainable higher performance.

> **Framework Vision: Placing the professional in charge**
>
> Our vision places the professional in charge of their own practices, supports them in identify where changes are needed, and empowers them to make those needed changes. Some fear this approach believing that project pressures may lead personnel to make poor decisions degrading rather than improving their practices. This is one of the reasons why separation of concerns is so important. Separating the essentials from specific practice extensions ensures the essentials are never lost as we make changes to improve and sustain our higher performance.

But how does an individual or an organization go about figuring out their own repeating specific weaknesses? In the next chapter we tackle this question and we explain a common measurement mistake that hinders many organizations in their attempts to locate their most valuable potential improvement areas. In Chapter Four we give you a little more help on where to look for your most valuable repeating weaknesses to work on.

# Chapter Two Summary Key Points

- When you just let people talk about their job you often hear key patterns that lead to a small set of repeating specific weaknesses that identify the areas where we can gain the greatest performance boost for the limited time and effort available.
- You can think of *repeating specific weaknesses* as an aid to help you find the right things to focus on for high value performance improvements.
- Repeating specific weaknesses have a critical quality of hurting us at just the wrong time so if we don't do something about them we will end up on that frustrating yo-yo getting better for a period of time, followed by degradation of performance.
- One of the best ways to keep people motivated and interested in their work throughout their career is to involve them in their own continuous improvement.

# Chapter Three – Common Measurement Mistakes

*"As data are aggregated they lose their context and their usefulness"* Don Wheeler

So how does an individual or an organization go about figuring out their own repeating specific weaknesses? The answer is you have to start gathering data and measuring, but you have to be careful how you go about doing this because your performance improvement potential is dependent on what you choose to measure, and if you measure the wrong things you are likely to never find the most valuable repeating weaknesses you need to resolve to achieve and sustain high performance.

## Common Organizational Measurement Mistakes

What many product focused organizations do is focus on defects identified during test and integration. However, defects are not isolated to test and integration. Some organizations have addressed this by requiring large amounts of information to be entered into a defect repository for later analysis. Typical information requested includes a categorization of the defect, how the defect was found, when the defect was found, when the defect should have been found, and what caused the defect [7]. Nevertheless, for various reasons these large repositories have failed to provide the promised high performance payback. One reason is because people are often in a hurry when entering defect information because they are under pressure to get the defect fixed, and therefore they don't take adequate time to ensure the information they are entering is accurate. Another reason is that this information is still general in nature and doesn't by itself provide insight into the kind of specific root causes we need to identify to help us achieve real meaningful and measurable performance improvement. So what should we do differently to avoid this common measurement mistake?

> ## Framework Vision: An agile and iterative approach to improvement
>
> As we move forward in the book you will learn more about our vision including how it can help to avoid the common measurement mistake described in this section by encouraging a more agile and iterative approach to improvement. This is not to say that we don't want to support traditional approaches as well. Our vision is one that does not dictate a given method or life cycle. It supports all life cycles including traditional waterfall approaches. However, our vision calls for an agile approach to improving your way of working regardless of your chosen development method. This can help avoid the common problems often observed related to large inaccurate data repositories as discussed in this section of the book. It also supports maintaining what is working well, while you focus on improving the specific areas that are hurting your performance today.

## Start With Your Current "As-is" Process

The first step to high value performance improvements– and to avoid the previously discussed common measurement mistake– is to start by capturing your current "as-is" process. This is not to be confused with what some people might think you should be doing. It is what you **are doing!** Capturing the current "as-is" process should take only a fraction of the amount of time often spent by many traditional process improvement working groups. The reason too much time is often spent on this activity is because most spend the majority of their time discussing and brainstorming how they think the organization should work rather than how it actually works. This is the wrong place to start—in business or in a personal improvement effort. If you limit the discussion to writing down what you are really doing today, and eliminate any discussions about how you think things should work, you will save a great deal of time and unnecessary resource expenditure, and most importantly you will have the correct basis for the next step.

Once you have captured your real *"as-is"* process you now have the basis to make the most effective decision on where to focus your performance improvement effort.

The next step is to identify the areas where improvement can provide the greatest payback for the least investment.

> ## Sidebar: Process improvement caution!
> While I always stress "as-is" first, "as-is" is only the first step. By itself you won't see performance improvement from capturing the "as-is". I have a client who said, "we are wasting time talking about improving the process. I don't have time to talk about more than what we do today." **Beware** of this scenario, as it will lead to no performance improvement.

> ## Framework Vision: "As-is" first thinking
> Our vision supports "as-is" first thinking. We believe everyone should start with where they currently are (e.g. "as-is") and incrementally change in small steps with each step focused on what is most important to do next.

This is an area that traditionally has been very difficult to identify because of varied views of value to the organization. As Geoff Colvin alluded to in his discussion of practice there are probably thousands of different things you could be thinking about when it comes to a golf swing.

> ## Framework Vision: Helping you decide what is most important next
> While each organization must locate their own unique repeating trouble spots, our vision includes a framework that can help organizations by guiding them to consider areas where trouble has most often been found in the past.

However, my experience from working with multiple organizations over many years indicates there are just a critical few that can lead to rapid and high value sustainable performance payback. This leads to the next fundamental.

## Fundamental Five:

You should always, as individuals, reduce the number of areas you are focusing on at any one time to between three and seven, ideally closer to three. Organizations can have more, but this should not increase the focus of individual performers.

## Sidebar: Rationale for 3 to 7

Millers Law argues that the number of objects an average person can hold in working memory is 7 plus or minus 2. Recent research demonstrates the correct number is probably closer to 3 or 4. [25] This idea of limiting what you are working on is supported by Personal Kanban [26]

## Framework Vision: Change incrementally

Our vision stresses that all changes to your way of working should be accomplished incrementally in small steps supporting fundamental five.

# Where to Look for Your Most Valuable Repeating Specific Weaknesses

As you are looking for your repeating specific weaknesses keep in mind they need to create real obstacles that are clearly hurting your organization today. This is important for a couple of reasons. First, it gives us the best chance for high value and visible payback today. Second, it will get the attention of project managers (or coaches, or sponsors) because if your effort can be shown to help solve current organizational weaknesses you will find it much easier to get their buy-in and support for your related performance improvement activities.

Use the following criteria to help locate repeating specific weaknesses in your organization:

- Often first appear as small almost imperceptible issues making them easy to ignore (discussed in more detail in Chapter Four)
- Tend to occur during times of performance stress
- Create clear obstacles to achieving your objectives

# A Case Study of a Common Measurement Mistake

To quantitatively manage and gain performance benefits data must be collected, analyzed, and then actions taken leading to measurable performance improvements. Figure 3-1 provides an example of the type of data often selected for quantitative management today in many organizations.

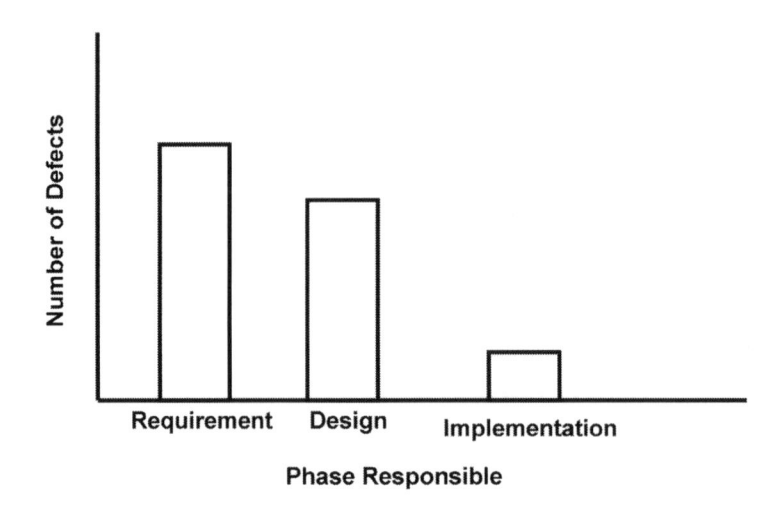

**Figure 3-1 Number of Defects By Phase Responsible**

Typical analysis of this type of data might go like this:

*Our data indicates we are making too many mistakes during the requirements and design phases. The data indicates we should have been finding these defects earlier during a peer review so we need to improve our peer review process of our designs and requirements. Our people also need better training so we should update our requirements and design training with the latest new techniques.*

As a result of this analysis people may be sent to training on the latest fad or hot topic in industry in the area the organization feels it needs improvement—in this case it could be Use Case training or Object-Oriented design, or training in the latest peer review process with a new peer review tool the company decides to invest in. On the surface this improvement sequence appears sound and many high maturity organizations have provided data that seems to indicate the process works. Reference Figure 3-2.

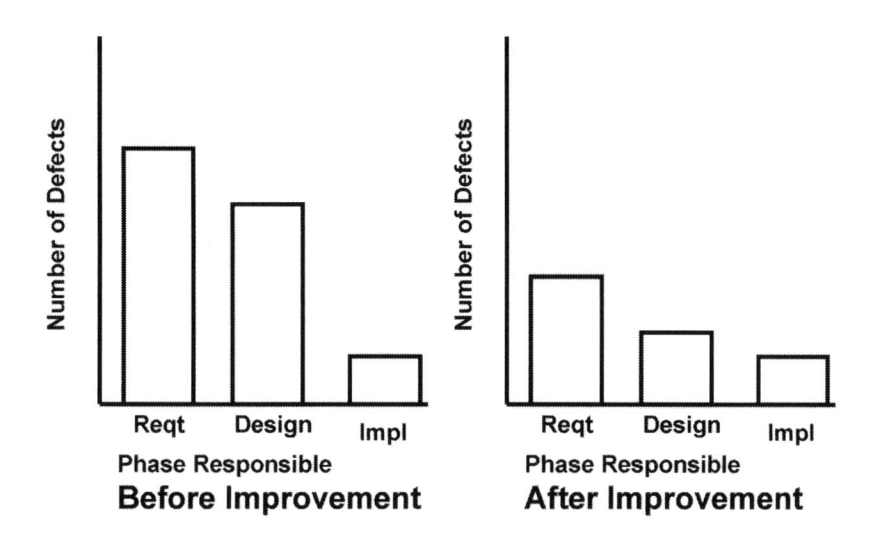

**Figure 3-2 Before and After Improvement**

But have they really made significant improvements in their performance? When you step back, and look at this objectively, especially from the customer side, it is often questionable. One customer said: *"I don't care about the CMMI rating of my contractors because I have never been able to see a difference in performance between a level 3 and a level 5 company."*

## Sidebar: How Lean Six Sigma can help

Lean Six Sigma [6] places a strong emphasis on defects being something the customer would view as not meeting their requirements. Too often organizations place their focus on the wrong type of defects leading to ineffective effort. Ask yourself, when considering the types of defects to quantitatively manage: *"Are these defects the type our customers would pay for us to resolve?"*

Interestingly, many have observed significant performance improvements as organizations move from maturity level 1 to level 3, but improvement beyond 3 has not been as clear inside many organizations.

Why is this the case? This brings us to another question:

*Are we really getting to the real root causes and the high value significant weaknesses inside organizations that are getting in the way of their people's performance?*

When you talk to people inside many organizations who have been through a typical improvement cycle, what we often hear is something like:

*I got told I had to attend some new training course that was suppose to help my performance, but it was basic stuff that I learned years ago and it had little to do with the issues I face each day on the job.*

So it appears many aren't getting to the real root causes that are holding people back from high performance. If your organization is going to invest in monitoring and managing a specific area you should do everything within your power to ensure it is an area that can lead to high value performance payback.

This means first digging deep—or as I call it, "following a thread" all the way through as we did in the Procurement Case discussed in the previous chapter.

This means basing your analysis on real facts based on real discussions with the real workers who know what is working and what isn't working within their project work environments. Too often we see areas selected for the wrong reasons, such as selection based on available data. Don't just select areas to monitor because they are easy to collect data against. This is a sure fire way to ensure your effort will not produce the performance improvements you desire.

# A Flaw Observed With Today's Common Analysis Approach

The flaw observed too often today is we put off analysis until after we have collected the data. This means we have selected our areas to quantitatively manage without doing adequate analysis, and this is where the problem begins. When we do this we are already doomed to not finding the real root of our problem because we aren't gathering the right information that can lead us to where the real problems lie.

By following a thread we can isolate an area where we know there is trouble in the organization. We may not yet know the solution(s), as was true in our procurement case, but we need to do enough investigation to be sure that we have identified an area that is causing the organization pain and therefore further quantitative measurements and analysis are warranted and likely to result in findings that can aid high value future performance improvements within a reasonable timeframe.

# Where to Look for Candidate Areas to Measure

Frequently the areas that should be monitored are not found in single processes, or practices in an organization because often the most troublesome problems cross department boundaries. This is where we frequently see processes breakdown under stress. If you are concerned that you may not know how to measure the areas you believe have weaknesses in your organization that could provide high value payback if solved, or if you are concerned that you will not be able to get a large sampling of data, pick up the book "How to Measure Anything" [27].

In this book author Douglas Hubbard points out how looking at just a few samples can tell us a great deal more than many realize, and that *"we may only need a very small number of samples to draw useful conclusions..."*. Remember the goal is to help your organization's performance, not gather a large amount of data.

## Fundamental Six:
When selecting areas to measure ensure you have done sufficient analysis (e.g. following real threads) to know you understand the real context and there is high likelihood of findings that will lead to improved performance in the reasonably near future.

Often when people are searching for areas of common pain (repeating weaknesses) in their organization they first notice symptoms. Because of this it is important that we continually review and refine the areas we are monitoring and the related measures. This will be discussed further in the next chapter where we dig deeper into the subject of how to locate your most valuable repeating specific weaknesses to focus on.

---

[9]Reference page 128 of "How to Measure Anything," by Douglas Hubbard

## Chapter Three Summary Key Points

- Capturing the current "as-is" process should take only a fraction of the amount of time typically spent by most process improvement working groups in many organizations.
- If you limit the discussion to writing down what you are doing today, and eliminate any discussions about how you think things should work, you will save a great deal of time and unnecessary resource expenditure.
- When looking for your repeating specific weaknesses keep in mind they need to create real obstacles that are clearly hurting your organization today.
- If your organization is going to invest in monitoring and managing a specific area you should do everything within your power to ensure it is an area that can lead to high value performance payback. This means first digging deep—or "following a thread" all the way through.
- Don't just select areas to monitor because they are easy to collect data against. The flaw is we put off analysis until after we have collected the data which means we have selected our areas to quantitatively manage without doing adequate analysis.
- Frequently the areas that should be monitored are not single processes, or practices in a single department. Often the trouble spots are processes that cross department boundaries.

# Chapter Four – Little Things That Aren't So Little

*"All great things are only a number of small things that have carefully been collected together." Anonymous*

In the last few chapters I motivated why identifying your repeating specific weaknesses is a critical first step toward sustaining higher performance, and I provided some initial assistance helping you locate yours. Your repeating specific weaknesses represent the primary obstacle between where you are and achieving the sustainable higher performance you desire. But finding your most valuable repeating weaknesses to work on– that is, the ones that can payback the most for the least effort– can be tricky and so in this chapter I provide a little more guidance to help you locate yours.

## Attaining High Value Improvements

What was it on my personal improvement project after so many years that helped me attain the high value improvements I desired? I first needed to understand my repeating specific weaknesses. I discovered them by taking notes each night capturing how I felt about my performance.

I had limited time available and I needed to focus that time where it could provide the greatest benefit. The second thing I did was to analyze the data I had gathered. This in turn led me to take some very specific actions. It is possible I may have actually known I had repeating specific weaknesses earlier in my life, but I wasn't consciously aware of them and I had never taken the time to determine if there was something I could do about them, and then do it.

Summarizing the steps I followed:

- *Collect specific data about my weaknesses*
- *Analyze the data*
- *Take specific actions to counter the repeating specific weaknesses identified*

## Collect, Analyze, Act: Fundamental to Successful Improvement

The steps collect, analyze, act are fundamental to all successful improvement approaches including CMMI, Lean Six Sigma and Agile Retrospectives [28][10]. By focusing on just my three key preparation weaknesses I was able to improve my performance more than I ever imagined given the limited amount of time I had to devote to it. This can work in business too, but in business there are often obstacles that must be overcome.

## Typical Business Stumbling Blocks

On the business side I have found that leaders in organizations are rarely surprised by my gap analysis findings exposing their most significant repeating weaknesses. In fact I almost always hear complete agreement with my findings, and I hear senior managers tell me they have known what I say is true for a long time.

This leads to an important question:

*Why is it that the problems still exist in many organizations when Senior Management has known about them for extended periods of time?*

In some cases the reason is that the right fixes to these known weaknesses appear so straightforward that they end up brushing the issue aside. Sometimes the answer just seems too simple. In other cases the organization may not know how to fix them. Because they don't know how to fix them they give up without trying and then they just keep coming back. Lets now look at a business case study of an organization that figured out what they needed to do to sustain their performance.

---

[10]Disciplined Agile Delivery (DAD), http://Ambysoft.com/books/dad.html, provides similar advice around metrics through an ongoing process goal called Improve Team Process and Environment. This can be accomplished through retrospectives.

# Business Case Study Demonstrating Sustained Performance

I have a client that understands that business objectives must come first and drive any measures collected and analyzed and any improvements. These are the essential practices to all effective performance improvement approaches.

What is interesting about this client is they are a large organization and they have complex products that often involve projects that go on for years and include subcontractors that operate in different parts of the world. After many years of trying different approaches to manage their projects they finally figured out the key for them to sustain high performance:

*Focus on just a handful of key measures and continually remind everyone in the organization, and their subcontractors, what the goal is and what was expected to be periodically reported and managed.*

The measures were nothing that would surprise you. They were the usual cost, schedule, staffing, issues, and risks. However, what was different was the data presented had to be real with objective back up. No one could get away with hand-waving their status reports, which had become common in previous years when they had experienced common cost and schedule overruns and general poor performance.

Another change this organization made was to centralize all improvement effort under one point of contact to ensure effective use of resources and to ensure any improvements were aligned with the business objectives. This organization had not formally achieved a high CMMI maturity rating, but interestingly they used the CMMI model level 4 and 5 practices in an informal way to help manage their process improvement activity.

This organization did not have a large process improvement budget. Therefore, they were careful to look at the cost benefit situation for each improvement idea they received. In previous years they had a much larger process improvement budget and it was not uncommon for expensive and long improvement studies and pilot projects to have been conducted. But the results of those efforts rarely seemed to find their way to the people who were faced with real project dilemmas.

Since their process improvement budget had been significantly reduced they started looking closer at suggestions from the practitioners on their currently active projects that included using open source free software development aids. Using open source software had never been a part of this organization's culture, but they decided it was time to try something different. They also gave their development teams more freedom to make changes to their processes directly on projects. Previously this had to be approved through a long and bureaucratic corporate process.

These changes created a new level of process interest and enthusiasm in the company. For the first time in many years project team members started believing the organization actually did care about improving their work environment. What is interesting about this case study is the approach that finally got people to believe the organization cared about them was a change that occurred due to a budget reduction in process improvement activities. The budget reduction forced the organization to look to its workforce for help which is where they should have been looking all along to get high value performance payback.

## What's Happening Today in CMMI Level 5 Organizations?

In the beginning of the CMMI for Development V1.3 Guidelines [2], it[11] states:

*"At this point, of the many organizations that have been evaluated at CMMI Level 5, too many have essentially stopped working on improvement. Their objective was to get to level 5 and they are there, so why should they keep improving? This is both an unfortunate and an unacceptable attitude... it is unacceptable because the essence of level 5 is continuous improvement."*

So if it is true, that Senior Management knows about key problems in many organizations, why are we not seeing continual improvements, especially in CMMI Level 5 organizations?

## Problem Isn't That We Aren't Trying

I believe the problem isn't that organizations aren't trying to get better. It isn't that they aren't expending resources at improvement initiatives, but rather that they are

---

[11]Reference page 7

not spending those resources on the right things and in the right way that can really make a difference in their organization. As a simple example, we should be placing more emphasis on the small changes that can help project performers today, and helping those performers learn to make these small changes correctly and consistently. Too often we think that small simple changes don't require effort, and because of this they aren't given the attention they deserve and thus the potentially valuable small changes frequently do not actually happen.

> ## Sidebar: Agile Retrospectives and small changes
> 
> With teams using an agile approach one of the purposes of the retrospective at the end of each development iteration is to identify small actionable improvements that can help the team perform better during the next iteration. But just identifying these potential improvements doesn't by itself ensure real performance gains are actually achieved. In organizations I have observed using agile retrospectives, while some are continually improving, many are not. If you are not measuring, you don't know for sure which group you fall into. [11, 28].

# Gaining High Value Performance Benefits from Small Changes

So what can we do to gain the high value potential performance benefits from small changes? First, it is important to recognize that there are usually numerous small things that could be selected, but only a few that may actually lead to something beneficial. To help select the beneficial small things ask yourself:

*What small things are hurting our performance the most and keeping us from achieving our objectives?*

Try to identify 3 or 4 candidate improvements and then narrow it down to just 1 or 2 that could help the most and are within your control to implement.

Second, you need to help your people learn how to take the right action at the right time to effectively implement the small change. Because the change seems small, don't be fooled into believing it will be easy to implement. This is the mistake that is too often made. We think because it sounds like something small it is easy and therefore requires little attention whereas in fact most small things that are truly hurting your organization's performance requires the kind of behavior change that takes practice for people to perform effectively on the job. Let me give you an example.

## A Small Change Requiring Practice To Master

Many years ago one of my first jobs after I started working independently was as a deputy project manager on a large U.S. Department of Defense project. Very early in the project it had become clear to me that no one in the organization was owning a key requirement for a support system. I started raising this as a risk, but it quickly became clear no one wanted to hear about it. This was because all of the project leaders had more work on their plate than they could handle, and didn't want to take on yet another responsibility. Since I wasn't getting any attention to my raising the risk, I stopped bringing the subject up at the periodic risk review meeting.

About two years later when the system was in final test and almost ready to be accepted the customer asked about the support system requirement. They said they would not sign off and provide final payments until this requirement was met. It ended up delaying final acceptance by four months. As soon as this impact became apparent, the project manager starting asking all the leaders under him why no one had been warning him about this risk.

I learned an important lesson on that project that might seem like a small change to make. Once you have a risk, until action is taken and it is mitigated it is still a risk.[12] But learning to continue to raise a risk and do it every week until someone pays attention and action is taken to mitigate it, is not always an easy thing to do. It is something I have learned I still have to remind myself of frequently because it is just too easy to fall into old habits.

I have observed this same pattern occurring multiple times since that experience, and

---

[12]DAD provides an example of a framework with built-in risk management. Refer to http://disciplinedagiledelivery.wordpress.com/2013/11/28/agile-risk-management/

I am always on the look out for it. I now know the signals, and I now know when to take action so this specific scenario doesn't happen again.

I have learned how to take the right actions at the right time by keeping constantly aware of the conditions where I could potentially make the same mistake again. This is not simple to implement because it requires practice to keep the conditions fresh in your mind. As long as I stay vigilant I won't repeat this pattern, but I also know it could return in an instant if I stop paying close attention.

## Helping Practitioners With Small Changes

With the way traditional processes have been written (e.g. sequence of steps) they are of little value in helping practitioners recognize situations that call for timely actions. These traditional sequential processes work well when everything goes according to the plan, but they fail to help individuals when they need help the most– recognizing a situation that calls for timely action and giving them the help they need to reason through that situation and make the best possible decision.

> ### Framework Vision: A thinking framework helping with small changes
>
> Our framework needs to be a "thinking framework" in the sense that it helps individuals and teams make better decisions related to small timely changes by providing the team with objective data they can use to support their decisions.

## Another Reason to Focus on Mastering Small Changes

There is another reason to focus on mastering small changes which is captured in our next fundamental.

## Fundamental Seven:

Some of the most significant impacts to performance start out as seemingly little things that we often fail to notice until it becomes too late to correct.

The most well known example of this fundamental can be found in Fred Brook's classic book, "The Mythical Man-Month" [29] where he asks:

*"How does a large software project get to be one year late?"*

Answer: *"One day at a time."* [13]

While many of us can relate to this answer, it has been my experience that we often fail to notice those seemingly little decisions that cause small slips to a schedule. I have observed this fundamental in my client organizations, and in my personal performance.

One of the reasons I believe this occurs is because we start to take success for granted and stop paying close attention. The next thing you know you are just *going-through-the-motions* without the proper thought, and going through the motions does not work.

Another reason is because sometimes we just get overwhelmed with too many little things and not enough time to get them all done. Then you are faced with having to make a hard decision on where to focus your limited time.

## Framework Vision: Helping practitioners make difficult decisions

Helping practitioners make tough choices where they don't have enough time to do everything may be the most valuable area where our thinking framework can benefit our teams.

---

[13] Fundamental Seven is also related to the popular notion of "signal-to-noise ratio" in a system. Informally this refers to the ratio of useful information to false or irrelevant data in a conversation. Refer to http://en.wikipedia.org/wiki/Signal-to-noise_ratio for more information.

Also, don't forget that it is your repeating trouble spots that keep you from sustaining whatever improvement you want to sustain. In other words, an important characteristic of repeating specific weaknesses is that by their nature they are *never completely resolved.*

> ## Glancing Forward
> 
> The fact that repeating specific weaknesses are never completely resolved is one reason why traditional defect resolution approaches that attempt to apply a single "quick fix" fail to provide sustainable performance. In Part II of this book we will discuss a better approach to handle these forever returning trouble spots.

## Personal Improvement Project Repeating Specific Weaknesses

My personal repeating specific weaknesses all occurred during my set up. First, I had a tendency to allow my left thumb to move to the top of the club. Second, my left shoulder would start opening up and pointing to the left of the target. Third, I would place the club face down behind the ball with it facing to the right of the target. The net affect of this misalignment was loss of distance and pulled shots to the left or sliced shots to the right. When my teacher would correct me it always felt odd and it always took a few days for this correction to feel natural. This leads to the next fundamental.

### Fundamental Eight:

> When you've been doing something wrong for an extended period, the right way may feel wrong for a period of time while you are adjusting to the change.

> **Framework Vision: Where the framework helps most**
>
> We envision the framework helping throughout the life cycle although its value may appear less visible to practitioners at certain times. For example, if the project runs smoothly, its value may appear to diminish as the project proceeds. If, on the other hand, the project starts to run into trouble, the value of the framework will quickly become more evident. This is because the framework is a monitoring aid. You can liken it to a good referee in a sporting event. In well played games good referees are often not noticed although they are still critical to the success of the event. When the trouble starts their value quickly becomes evident.

This is true for individuals and organizations, which is part of the reason why both individuals and organizations revert to old behaviors during times of stress. Fear leads us toward comfort, not toward what is necessarily best for performance. Most organizations know the repeating patterns they suffer from, but they don't know how to stop. Examples of common patterns mentioned earlier include organizations that drive solutions to a crisis from the management side while failing to provide leadership and guidance deep in the organization during the crisis, and organizations that drive solutions from the technical side while losing cost and schedule accountability.

Similar to the story earlier in this chapter on a small change requiring practice to master, on the personal improvement project I learned to be on the look out for a repeating pattern to be ready to take the right action at the right time. I put continuous checks in place to ensure I was avoiding my known trouble spots by catching them quickly and resolving them before they could hurt my performance. In the next chapter we discuss the types of checks I put in place to accomplish this, and why these checks are not sufficient to sustain high performance.

# Chapter Four Summary Key Points

- Your repeating specific weaknesses represent the primary obstacle between where you are and achieving the sustainable higher performance you desire.
- The first steps to eliminating your repeating specific weaknesses:
    - Collect specific data about my weaknesses
    - Analyze the data
    - Take specific actions to counter the repeating specific weaknesses identified
- Business objectives must come first and drive any process improvements and any measures collected and analyzed.
- When considering areas of improvement we should be placing more emphasis on the small changes that can help projects today.
- The problem in many organizations is that they aren't spending their improvement dollars on the right things at the right time.
- Some of the most significant impacts to performance start out as seemingly "little things" that we often fail to notice until it becomes too late to correct.

# Chapter Five – First Level Checkpoints: Necessary, But Not Sufficient

*"Everything Should Be Made As Simple As Possible, But Not Simpler"* Albert Einstein

In this chapter I first explain a simple technique I used to help keep my personal repeating specific weaknesses in check, and how this technique relates to performance objectives. Then we explore drawbacks of this technique and I explain why it isn't sufficient to help you achieve and sustain the high value payback you seek.

I know that fundamentally I have a good golf swing and I didn't want to get in the way of allowing my natural swing to happen. But at the same time I knew I had to find a way to stop my key weaknesses from continually repeating right when I needed my performance to be at its best.

By using my pre-shot routine (explained in the next paragraph) with conscious checks before each shot– which I refer to as first level checkpoints– I was able to make small corrections before each shot keeping my weaknesses in check while not intruding on my actual swing.

> **Framework Vision: Non-intrusive approach**
>
> Our framework vision is one that does not try to change what is already working well for you. It gives your team reminders of where they need to focus their attention given where they currently are. When your team realizes they are having trouble in a particular area, the framework can give you extra guidance helping you find the right solution for just the specific area you currently need help with. But the framework will not require you to solve your challenge in any particular way. This will help your team improve without forcing them to change what is already working well.

## How I Countered My Repeating Specific Weaknesses

I created a *pre-shot routine* where I deliberately looked at my grip to see if my thumb was sliding. Over the next few weeks I caught myself a number of times trying to move that thumb back up on top of the golf club shaft. I don't know what makes me do it. But I know now how to correct it before it does much damage to my performance. I added a similar pre-shot checkpoint for the other two known weaknesses that are common with my swing. This *pre-shot set-up with conscious checks* appeared to have gotten this issue under control. But I have to keep monitoring this trouble spot because it is a repeating weakness. In other words, it is a risk to my swing and if I stop monitoring it, I know it will come back.

## Definition and Characteristics of First Level Checkpoints

We define first level checkpoints to be:

*Checks that you can execute quickly to test if you are doing something right. ( Not every aspect of it, just those few that you need to watch).*

A criteria I have found helpful in identifying effective first level checkpoints to counter repeating specific weaknesses includes:

- Non-intrusive
- Support rapid feedback
- Support continual small corrections

> **Framework Vision: Focus on essentials and provide easy to use aids**
>
> Our framework will provide simple easy to use checks as well as some that require more practice to apply correctly. These checks will help teams assess progress and health in a consistent agreed to way.

Many organizations use first level checks throughout their business today. For example, just before you start a meeting in many organizations someone checks to see if the key people required to address the intended subject are in attendance.

As another example, before software is released in many organizations there is a check to make sure the assigned personnel have conducted the required design, test and review steps. Checks that ensure these types of best practices have occurred can usually be accomplished rapidly, and appropriate actions initiated on the spot when the situation dictates.

Lets now discuss how first level checkpoints relate to performance objectives and measures.

# Performance Objectives, Measures and First Level Checkpoints

Key to any improvement effort is to first make sure you understand your performance objectives, and then determine how you will measure yourself to determine if you have succeeded. Without objective feedback you have no reference from which to determine if you need a course correction, or if your action is successful [30]. My performance didn't need to be perfect to achieve my golf objective, but I needed a way to manage my performance to ensure it stayed within acceptable limits.

When I was a teenager I played close to par golf (shoot in low 70's on many par 72 golf courses). At 59 years old given the amount I played I would be satisfied to play well enough to have a reasonably good chance to break 80 each day. That was my measurable objective for the project. A more specific measure that I derived from that objective was "thumb position." This was something I could monitor quickly by visually checking prior to each shot and make a small correction on the spot, if necessary.

Another observation that I found in my notes about my performance was my tendency to bring the club head too far inside on the backswing. Chris Demarco, a touring professional has a habit during set up of taking the golf club back half way and then looking at it. I have found using a similar practice swing and checkpoint helps to ensure that I break my wrists up soon enough keeping the club on plane while starting the backswing. I can make this check quickly just before each shot, and make a small correction on the spot, if necessary. As a result I have also added this checkpoint during my pre-shot routine.

> ## Framework Vision: Adding or extending to the essential things we work with
>
> Our framework vision includes the capability to add other important things you decide are important to monitor and progress within your own business environment.

# How First Level Checkpoints Work

The checkpoints I have described so far I refer to as first level because they work by sensing quickly common repeating weaknesses and providing feedback so they can rapidly be corrected. First level checkpoints can improve your performance by improving your consistency. But my first level checkpoints didn't solve all my performance problems. They actually did a very good job to help me achieve my immediate goal on the personal improvement project. I played very well and consistently for the full marathon golf week. But I found that after that week I had trouble maintaining the level of play I had attained. There are multiple issues to be aware of with first level checkpoints which demonstrate why we need something beyond them to attain and sustain the high value performance payback we seek.

# Why First Level Checkpoints Aren't Enough

While first level checkpoints can help, they are not sufficient and you have to be careful how you implement them. First, you don't want too many, and you don't want them to get too complex. This is because if you just keep adding more first level checkpoints you can actually degrade performance. This is what too many organizations try to do. Unfortunately, as we add more first level checks people tend to get overwhelmed. When you try to think about too many things at once your performance is certain to degrade.

> **Framework Vision: A Caution about the framework**
>
> We must be careful not to let our framework become burdensome to use, and we must be careful to ensure the framework will apply to all software engineering endeavors. Because our goal is to help everyone with essentials for all successful software engineering endeavors, it cannot possibly give you everything you need for the success of your specific endeavor. We encourage teams to add what is important to monitor in their specific situation, but we caution you to stay alert to the degraded performance that often occurs when practitioners are asked to recall too many things at the same time.

This actually started to happen on my personal improvement project. As I found new weaknesses I tried adding more checkpoints, but as I did my original alignment checkpoints started to fail because I stopped paying close attention to them. Individuals can only focus on a small number of items at a time so it is critical that we pick the most important ones to aid performance.

Second, while each first level checkpoint represents something that can help you achieve your performance objective, all of the first level checks taken together will not ensure you achieve the high value improvement payback you seek.

As an example in business, checking to make sure the key people are in attendance at a meeting will not ensure those people achieve the goal of the meeting. Similarly, checking that software has gone through all the required design, test and review steps will not ensure the software achieves its goal of satisfying customers when it is deployed.

Furthermore, depending on your specific situation some first level checks may be critical to achieving your objective, while others may take on far less significance. First level checks are based on past experience, but they will never be perfect because future situations don't always reflect the past.[14]

# First Level Checkpoints in the Business World: Quality Assurance

Many organizations have quality departments that employ extensive quality checklists. The CMMI model also encourages quality assurance checks to provide management with objective insight [2][15].

Quality groups usually do not check everything. Due to staffing constraints most conduct their quality assessments using a sampling approach. One complaint often heard about many quality organizations is that their effort does not contribute to high value performance because too much of the focus is on minor defects such as documentation formatting and punctuation. One approach organizations may want to consider is to develop quality checklists more closely aligned with key pain points that have hurt their performance repeatedly in the past.

---

[14]There is a useful analogy that can be made between first level checkpoints and what is referred to as specific and generic practices in the CMMI model. These practices are expected, but they are no assurance of achieving the CMMI Goals and it is the goals which are most important to your performance. This fact is also a motivation for the use of patterns discussed later in the book.

[15]Reference specifically the Product and Process Quality Assurance (PPQA) process area within the CMMI model.

> **Framework Vision: Value**
>
> By using our framework organizations will not need to start from ground zero when developing their quality assurance approach. The framework will give you a common widely accepted base from which to start. You can then add items specific to your business and your specific organizational weaknesses that you want to monitor more closely. If you already have your own quality approach, then you can use the framework as an independent cross-check. This is an example that demonstrates one way organizations will be able to take advantage of the framework in a complementary way to whatever they are currently doing.

# An Example of Improving Your Quality Assurance Checklists

In one of my client organizations when I asked a Quality Assurance representative if she ever raised risks, she replied:

*"Absolutely. Just last week the customer asked for some added functionality in the next release which is only a week away. A developer said, 'sure, we can do that', but I immediately objected to the program manager because we didn't have adequate time to run all the regression tests."*

This is a simple example of a Quality Assurance person sensing a past pattern in the organization and making a decision to raise the risk to the program manager. This type of check starts to move us beyond the simple first level checks discussed in this Chapter. It is starting to get at the real goal we seek– that is high quality software. This type of check will be discussed further in the next few chapters where we talk about second level checkpoints.

Consider the following when developing your quality check guidelines:

- Focus on items where your organization has historically had performance weaknesses
- Keep checks to a small number by removing out-dated and non-value-added items

## Common Mistakes with First Level Checkpoints

With one of my clients when we instituted first level checkpoints with some new processes we were deploying, the organization didn't fully understand the drawbacks of making first level checkpoints too complex. As a result they tried to look at too many things. When it became clear they couldn't possibly do everything their checks called for they became overwhelmed and reverted to their old behavior.

It is common for people when they first start to learn to play golf to fall into a similar trap. The golf swing is very complex, and it is easy to start thinking about too many things at once as Geoff Colvin alluded to when describing how he practiced golf.

One problem with first level checkpoints is that they are susceptible to our just "*going through the motions*" when executing them. This is because of their simplicity which is good because we can execute the checks quickly, but over time it becomes difficult to maintain proper attention on these simple checks which causes their effectiveness to degrade.

Even though they are simple, first level checkpoints can also be misapplied. On my personal improvement project over time I started misinterpreting what it meant to place the club head down squarely behind the ball. I found myself asking, "square to what?" I would look at the club head and think I was laying it down square, but since I had no *objective* reference I would gradually fall into my old habit of laying it down with an open club face.[16]

In the business world the analogous situation that often occurs is when quality groups just check for the existence of certain artifacts, rather than checking to see if the real

---

[16] In golf an "open clubface" means the face of the club is pointing to the right of the target.

intended goal was achieved. For example, the existence of a requirements document is no assurance that key stakeholders are in agreement with those requirements, nor is it any assurance that the requirements are clear, complete and consistent.

The higher levels of performance we seek can't be achieved through fundamental checks alone. Observing fundamental patterns that reoccur is necessary and the good news is straightforward fixes can often be applied easily and rapidly. However, we need something more powerful to attain and sustain the high value performance payback we seek.

# Examples of 1st Level Checks/ Where Business Needs More Help

Examples in the business world of first level checkpoints include verifying the right people attend a design review, and ensuring all requirements are addressed in a design approach. But on the business side and those seeking personal higher performance levels we need to move beyond just sensing the easy to sense and easy to fix weaknesses.

In business there exist *grey* areas where the proper course of action requires deeper reflection and analysis. Examples include:

- decisions related to priorities of work
- decisions related to changing the responsibilities of individuals
- decisions related to speaking to a performer or teammate about his/her performance
- decisions related to talking to a customer about an ambiguous or vague requirement
- decisions related to taking action to mitigate a risk

If we are to continually raise our performance to the higher levels we seek we need more than what the simple first level checks can provide. Not all situations, especially those faced during times of stress, have simple black and white answers. It is during these times of stress where most often we fall back. But it is also during these times of stress where often our greatest opportunities lie and where our sustained higher levels of performance can benefit us most.

We need to find ways to detect the areas that cannot be addressed with simple fixes.

We need to learn how to sense those hard to sense areas right at the moment when action should be initiated, and we need to learn how to help ourselves select the proper action. Techniques to help us sense these hard to sense areas and more powerful techniques to help us take the right actions at just the right time are the subject of Part II of this book.

# Chapter Five Summary Key Points

- The simplest type of check you can put in place to help you avoid your repeating weaknesses at the right time are first level checkpoints. Their characteristics include:
    - Non-intrusive
    - Support rapid feedback
    - Support continual small corrections
- Key to any improvement effort is to first make sure you understand your objective, and then determine how you will measure yourself to determine if you have succeeded.
- First level checks work by sensing quickly common repeating weaknesses and providing feedback so they can rapidly be corrected.
- First level checks are necessary, but they are not sufficient for sustained high performance.
- To gain more value from your quality checks, consider keeping your quality checklists more closely aligned with your high value pain points in your organization.

# PART II Beyond Fundamentals

# Chapter Six: The Path to Higher Performance

*"I used to work in an organization that had a quality program. Once a year there would be a flutter of activity to get the documentation up to speed before the reassessment. The rest of the year they ignored it. Its not the way it should be. If we are doing it to check a box, then its not of value. If doing it to help make us better, then I support it." Program Manager in Organization Starting Improvement Effort.*

Let me begin this part of the book with a little background on where I came up with the idea I am going to present in this chapter and the different kind of practice we are going to discuss. Many years ago I was conducting a project management workshop for a client and I was teaching the fundamentals of planning. During a break one of the participants came up to me and said:

*"This all sounds good, but I don't know how I can apply these principles in the fog-of-war of my real project."*

As we talked further I began to understand his dilemma. I was giving him a theory of planning, and he was telling me he didn't see how he could use the techniques I was describing in his real work environment because of the conditions he was being forced to work under. For example, in this case he was not being given enough time and budget to actually follow the planning techniques I was describing. What I realized from this experience was that this student needed more than an understanding of the principles of planning. He needed help connecting the theory of planning to the actual environment in which he was being asked to perform. He was struggling to make this connection on his own.

This brief interchange led to a deeper discussion into his specific project situation which in turn led me to modify the way I was presenting some of the techniques in the workshop to help more people learn how to achieve the real intent of planning (e.g. the goal) and be able to actually use what they learned back in the real world of their projects.

Earlier in this book I talked about how we initiated what I referred to as *"Sustainment Training"* sessions with one of my clients. What often occurred during those sustainment training sessions was similar to the deeper discussions we had on planning in this workshop in that we were taking the theory of planning a step further to demonstrate *how to* apply it to *specific* real project conditions. [31, 32]

The first point I want to emphasize about the value of the second level checkpoints that I am going to describe in this chapter over the simpler first level checkpoints is that they are better at assessing **how well** you are doing toward achieving your goal given the specific situation you face.

Following is a quick business example to help you get the idea. As discussed in the last chapter a first level check for a review of a requirements specification could be to check that we have the right people at the requirements review meeting. But the goal of the review is to ensure we end up with a requirements specification that is clear, complete and the key stakeholders agree to. Effective second level checks will help you more with ensuring you achieve the goal than first level checks. Thus, the second level checkpoints help you more with performance.

Now let us return to the personal improvement project to see how I discovered my own second level checkpoints and how they work, and then we will look closer at the procurement case study we began discussing in Chapter Two.

## Starting to Discover Better Performance Checklists

After my personal improvement project in 2008 I wanted to see if I could sustain the same level of performance I played with my friend and possibly even improve my performance beyond that level. So in 2009 I continued taking notes, and evaluating my checkpoints to try to continue improving my performance. Following are notes from the summer of 2009 personal improvement effort:

*"I don't know why I seem to keep learning the same lessons. Today, once again I caught myself falling into one of my old repeating patterns. But what was good was that I caught it faster than I used to catch it, and I was able to correct it sooner. I caught it by staying alert to the results of each shot. I could see the ball going shorter and starting to fade, and this was an immediate trigger to me that I was falling*

*into one of my common repeating specific weaknesses. I am starting to realize the power of this **objective external** check that I was getting by observing the results of each shot. It helped me to sense sooner when my repeating pattern was returning. I find it is very difficult to assess myself accurately by just using my normal first level checkpoints alone. I believe this is because my vision of what I am doing becomes clouded. I at times think I am checking my grip, stance, and club face, but still the recurring problem gradually happens. I need more objective ways to measure myself to maintain my performance."*

> ## Sidebar: General
>
> It should be noted that my motivation for continuing the personal improvement project in 2009 was mostly due to my business interest in performance improvement rather than any personal interest at this point in my life with getting better at golf.

## The Wrong Way to Conduct Root Cause Analysis

Note that in the situation described I was finding that same problem reoccurring. Collecting more data wasn't giving me new information, but what it was doing was confirming that my resolution to the problem was not effectively working. Doug Hubbard, in his book "How to Measure Anything" tells us that often we can learn far more than we realize with just a few samples of data, and that *"we may only need a very small number of samples to draw useful conclusions..."* [27]

I was actually able to identify my three primary repeating weaknesses within the first couple of weeks of my personal improvement project back in 2008. I then put checks in place that seemed to fix the problem. But over time I found the simple first level checks that I had instituted did not continue to work so well. I kept taking more notes thinking this would lead to the answer. But all it led to was more notes that told me what I already knew, and what it didn't tell me was how to permanently fix this problem.

What we must ask now is the question:

*Is it possible my three weaknesses are just symptoms of a deeper problem with my performance, and if I could locate that deeper problem at the root then maybe I could find a permanent fix that would allow me to sustain my performance much longer into the future?*

I have observed similar symptoms being managed, rather than getting to the root, in many of my client organizations. I have also observed many high CMMI mature organizations facing similar situations and handling the issue as I did. That is, collecting more of the same data which only tells you the same thing you already know. You still have the same repeating problem because you haven't really gotten to the root of the issue. As an example, many organizations collect the same data year after year related to defects– as we saw in the example in Chapter Three– without thinking about how they are using that data and what other data they might collect to help them prevent the same type of defects from occurring again in the future.

## Fundamental Nine:

Collecting more and more samples of the same data won't help an organization improve or sustain higher performance.[17]

Given this situation, what do you have to do to get to the real root cause that can then help you put the kind of resolution in place that truly affects your performance?

# Challenges Related to Effective Root Cause Analysis

A tip in the CMMI Guidelines book [2] with respect to Quantitative Project Management (QPM)[18] tells us, "This practice has a dynamic aspect to it. Although many measures and analytic techniques can be identified in advance, and thus the

---

[17] It is worth pointing out that while this fundamental is true, collecting more samples might get them to recognize that they have a real problem. Many organizations want a fair bit of evidence before they will act. What is enough evidence for an outside expert may be insufficient for insiders that have been there for years.

[18] Quantitative Project Management (QPM) is a Level 4 process area in the CMMI model.

project team is prepared to apply these and interpret results, specific situations may arise during the project that may require *more in-depth analysis to diagnose the situation and evaluate potential resolutions...*"

I have highlighted the words, "more in-depth analysis to diagnose the situation and evaluate potential resolutions..." This is exactly what I found I had to do with a number of business clients and on the personal improvement project.

As an example, now let us go back to the case study where we had isolated a problem area in the procurement organization, but we hadn't yet established a clear resolution to the problem. If you recall, the difficulty we faced was the fact that there were a combination of factors involved. First, we had a new employee that needed some coaching on how to handle a very specific situation in regard to missing data on a hardware requisition form. That fix helped the immediate problem, but on further analysis we realized there were other situations that could occur, such as other missing data items that would require the new procurement specialist to take a different action to resolve. What we realized after further analysis was there were so many possibilities that if we tried to describe what to do for each one it would require a manual of over 100 pages.

Even if we could get the funding (which was doubtful) to develop such a manual it probably still would not account for every possible case, and even if it did we still had the possibility that a new procurement specialist might not know where to look in such a complex manual to find what he needed right when he need it. To make matters worse in this organization the procurement specialists were all overworked and under a lot of stress because they were constantly being asked to process requisitions on short notice, and so they didn't have time to be reading 100 page manuals even if we created one. At this point the likelihood of isolating the root cause and resolving it was not looking good.

We realized we didn't have the answer, but we also realized that continuing to collect the same information was only going to confirm what we already knew. If we were to get to a sound resolution we needed to look at the data we had from a different perspective and possibly collect some different information that could shed more light on what was happening. This leads us back to the tip in QPM. We needed to refine the data we were collecting and, possibly, refine our resolution and then monitor to see the effect on the procurement specialists performance.

*Resolving a problem takes more than just identifying what appears to be the root of the problem. Often, what appears to be the root, especially with repeating stubborn problems, is not.*

You must continually refine your measures, your objectives, and your resolution until you really understand what is going on, and your results confirm it.

## Discovering Second Level "Feel" Checkpoints

On the personal improvement project I was continually refining my measures, such as adding more objective checks (e.g. watching the results of each shot) to better understand what was happening. I was also not just collecting data. I was continually refining my resolution to the problem based on the feedback which indicated how well the previous resolution was working. This helped me sense (or "feel") how well I was performing my fundamental first level checkpoints. This was a rapid feedback and refinement system, and through this process I seemed to be closing in on an appropriate resolution, although the root cause was still not clear to me. This led to what I refer to as second level (or "*feel*") checkpoints.

We define second level checkpoints to be:

*Checks that assess whether you are achieving the intended result.*

Second level checks also have a rapid feedback loop (like first level checks) that ensures timely actions are taken. Refer to Figure 6-1.

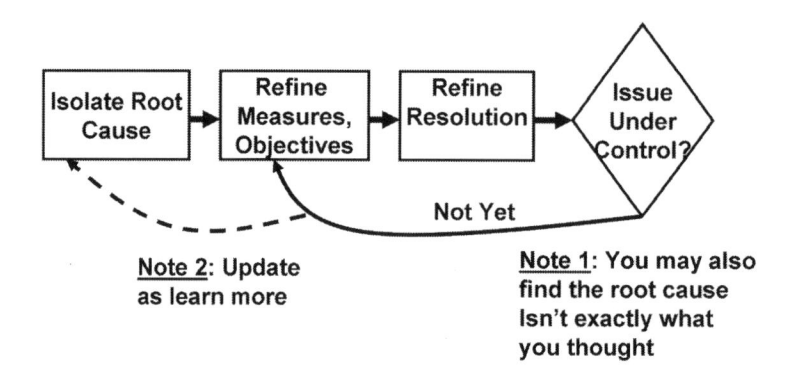

Figure 6-1 Path to Real Problem Resolution

# Motivating the Need for a Different Kind of Practice

Let me now provide an example of how I began to learn to use second level checkpoints effectively, and how it led me to understand more about the different kind of practice I needed to be doing. Then I will explain how this translates to the business world.

On the personal improvement project where I monitor the ***results*** of each shot, this is not a simple first level checkpoint where I can immediately take a single clear action based on the checkpoint feedback. When you are executing this type of checkpoint you will often find yourself in a high stress environment at the moment the feedback from the second level check becomes available. In my case it would be easy to just get angry when I observe the golf shot going short and fading. When this happens I tend to get irritated.

In the business world this translates to common situations that occur with daily interactions between teammates and/or managers. For example, how do software developers respond when trying to interpret an unclear requirement and they can't get their customer to take the time to talk to them? Or how do they respond when they haven't completed all their planned tests, but their manager is pressuring them to get their software released? What if they see a new risk with their design approach, but their colleagues are too busy to listen or help?

In each of these situations do they know the options they have, and which options have worked best in the past? In the case of a new design risk they could call for a special review before proceeding, but before making such a decision they need to think-through the consequences that exist on both sides of their decision.

For example, their teammates could have plenty of work on their own plate and everyone might be driving to meet an aggressive schedule. So calling a special review could impact the planned work schedule of colleagues. It might be possible to minimize this impact by limiting the review to just one or two key colleagues and focus the discussion on just the area of the design where the risk exists. But is there risk to dependent systems that might be missed with this approach? Note here the options, and consequences.

This is part of the reason why second level checks require more practice to master, and why this practice needs to occur under similar conditions to what you will face on a real project. Practitioners must teach themselves to produce the proper response in that momentary stressful situation.

It takes time to learn to respond properly under the real conditions you face in the heat of the battle at work or in a competitive personal event. Many of the situations faced in the real world don't come with simple answers. You usually have options and you need to learn to recognize rapidly the conditions under which each option provides the optimum solution. The existence of options and consequences to think-through is what distinguishes second level checkpoints from the simpler first level checks.

Let me now give another example of a second level checkpoint in business to help you appreciate the difference.

## Example of Second Level Checkpoint In Business: Big Picture Feel

I have found that an effective second level checkpoint in business is to keep what I refer to as a *Big Picture Feel* of the project plan continually in one's mind. Let me explain more how this can help in business. I had a client I was helping and I had drawn the conclusion that they weren't following their plan. They were new to planning and were clearly heading off doing unplanned work and it was keeping them from achieving their project goal. My client disagreed when I brought this to his attention. He said that he did plan and that he followed his plan.

He was convinced that because he had finished his project on schedule that he must have followed his plan. It was a software project and because they got the code done, that was what he pointed to as the proof. But because I had the key points to his plan in my head (the *big picture feel*) I was able to quickly counter with, "No. Your plan called for more than getting the code running. It called for requirements and design reviews, and it called for requirements, design and test artifacts to be produced and approved and controlled. But those artifacts have not been completed and reviewed as the plan called for. And you need those because it was part of the goal to build a system that could be maintained after the original software developers are gone."

> ## Framework Vision: "Big Picture Feel"
>
> The state of work products alone (e.g. requirements and design documents) cannot capture the true state of an endeavor's progress and health. Our framework vision is one that will provide critical indicators helping practitioners keep a "big picture feel" of their endeavor's true state clear in their head even as they must focus on their daily tasks. It will also help them ask the right questions leading them to consider options and consequences.

Because I knew the key points to the plan in my head in a big picture view I could respond with a clear answer under stressful conditions. But it took practice to learn to do this. You can't just write a plan and put it on the shelf and expect to have a big picture feel for that plan. I teach this technique to project leaders so they too can respond rapidly with the right decision in the heat of battle. But I also let them know they will need to practice to be able to perform this technique effectively when they need it most.

Let me explain more about what happened with my client. That client didn't have previous experience writing a project plan. He was developing his first project plan based on what he had learned in a traditional project management class he had taken. The problem was that the artifacts he was planning to produce didn't fit the schedule he had to meet, nor did it fit with the skills of the people on his project who would have to produce those planned artifacts. Unfortunately he had failed to adequately consider how to scale his planned products to fit the project cost, schedule, and personnel constraints. The second level checks, such as big picture feel, won't give you the "how to" specifics, but they will provide an early signal when people need more help.

When it became clear to the workers that they couldn't execute the project plan they just reverted to their old habits of just focusing on getting the code to run. What was most interesting in this case was that management didn't even realize what the team had done. They actually thought they had followed the plan. The root cause of this common repeating pattern– which I have observed in other organizations as well– is the lack of a big picture feel for the developed plan. The leaders didn't believe in the plan they had developed. They had essentially gone through the motions of planning, but hadn't adequately considered whether or not the plan was actually executable given their real project constraints, and if it truly represented the goal they were seeking. As a result no one was really checking to determine if the plan was being followed.

> **Framework Vision: Helping your team stay focused for success**
>
> Although our framework vision doesn't use the terms first and second level checks, it does include a similar idea. Our framework provides checks to help you stay on course that are easy to use, and it includes checks that take more practice to apply correctly because they lead to multiple options to consider. At any point in time the framework can help you assess where you are and guide you in making your next decision helping your team stay focused for success.

You have to keep your eye on the real target that you want to hit. This is true in any endeavor where you want to sustain high performance– personal or professional. The scenario described is not uncommon in organizations that are new to planning. To learn to plan effectively takes more than just going-through-the-motions of planning. This is where practice using second level checkpoints can help an organization ensure they are getting the high value performance improvements they desire.

In the next chapter we dig deeper into second level checkpoints helping you understand their essential characteristics.

# Chapter Six Summary Key Points

- Collecting more and more samples of the same data won't help you improve or sustain higher performance.
- Second level checkpoints are better than first level checkpoints at assessing how well you are doing with respect to your goal and the specific situation you face and therefore they help you more with performance.
- Often what appears to be the root of the problem isn't. You must continually refine your measures, your objectives, and your resolution until you really understand what is going on, and your results confirm it.

# Chapter Seven: The Essentials of Second Level Checkpoints

*"It is better to be approximately right than precisely wrong."* Warren Buffett

My goal in this chapter is to help you understand second level checkpoints a little better, and to help you understand why they require more practice than the simpler first level checks.

Let me start this chapter by returning to the procurement case discussed in previous chapters. Once we realized how complex the problem was becoming as we considered all the possible scenarios we realized we needed to step back and look at the problem from a different perspective. That different perspective turned out to be a bigger picture view. What we eventually realized was that while there were literally hundreds of different possible scenarios for combinations of missing data, or inaccurate data, on the requisition form, these scenarios could all be categorized so they fell into five different types (or patterns) where each type required a specific action by the procurement specialist. Refer to table below for the five categories.

| Category Type (or pattern) | Action |
|---|---|
| 1 Missing data, Specialist knows how to find data | Fill in data |
| 2 Missing data, specialist doesn't know how to find data | Contact submitter, place req on hold |
| 3 Part not available, but replacement part known | Order part |
| 4 Part not available, replacement part unknown | Contact submitter, place req on hold |
| 5 Other situations | Contact supervisor, ask for help |

We then created a simple table similar to the one displayed, and the procurement specialist hung a printout of the table on his wall where he could easily reference it as he worked. We also suggested to the procurement specialist that whenever he processes a category 2, 4 or 5 he log in his records the date the requisition went on hold. Over time and with practice procurement specialists learned to keep the 5 cases in their heads so they could respond rapidly and correctly to common situations even on a busy stressful day. Eventually this system was partially automated with periodic alerts of requisitions that had been on hold longer than a week past their due dates. Once the system became institutionalized the problem of late hardware orders was measurably improved.[19] [33]

What we learned from the procurement case leads to the next fundamental:

## Fundamental Ten:

Often the best path to high value performance improvement, especially when you have a limited process improvement budget, is to spend less time collecting data, and more time analyzing what data you have collected, and then using the results of that analysis to keep refining your resolution and measurements to ensure you are moving in the right direction.

## Framework Vision: Method independent

Our framework vision is method independent, supporting whatever method, or approach you choose. However, the framework should strongly encourage an incremental approach to improve your practices over time to ensure you keep moving in the right direction.

# Simplifying Complexity Through Patterns

In the procurement case previously described it is important to note that while we were able to simplify a complex problem it still required the procurement specialist to think about each case, properly categorize it, and then take the appropriate action as

---

[19]How we solved the problem in the procurement case study has similarities to an industry initiative referred to as Adaptive Case Management.

specified in the table. The five types can be viewed as scenarios, or common patterns. When we analyzed the data we considered making more types, but we decided 5 cases was the right number to help the procurement specialist perform at his best.[20] Note here that we are keeping the number of items the procurement specialist must recall between 3 and 7. This is a very simple example of a second level check. I refer to this as second level because it requires more than a simple yes/no answer. It requires that options be considered and a decision be made.

## How Senior Management Can Help Practitioners Make Better Decisions

Douglas Hubbard tells us:

*"If we can't identify what decisions would be affected by a proposed measurement and how that measurement could change them, then the measurement simply has no value."*

When I asked what decision would be affected by the data they were planning to collect in one client organization, the answer I received was, "we've only been asked to collect the data. We have no authority to make a decision, or change the process." This is an area where Senior Management can make a difference.

*Don't ask your people to collect data. Empower the team to solve specific performance problems.*

This will lead them to collect the right data they need and immediately use it to improve performance in your organization. In Chapter Eleven two scenarios are shared that further motivate changes organizations can make in support of increased practitioner decision-making while maintaining appropriate discipline.

---

[20]For more information related to assisting decision-making by simplifying complexity refer to Snowden's Cynefin Framework, http://www.mpiweb.org/CMS/uploadedFiles/Article%20for%20Marketing%20-%20Mary%20Boone.pdf

## How Second Level Checkpoints Differ from First Level Checkpoints

I want to emphasize that, like the first level checks, the second level checks are not just data collection. They include rapid analysis, decisions and corrective actions, as dictated by the situation. But the decisions are not as straightforward as those with first level checks.

> **Sidebar: Second level checkpoints & goals**
>
> Some readers may think the second level checkpoints are goals for which first level checkpoints are a means to achieve. This is not correct, but it is understandable how this conclusion could be reached. The second level checkpoints do relate more closely to the goal. But their power comes from the fact that they are themselves a means to achieve the goal and provide a distinct value beyond the first level checks alone.

I refer to them as "*feel*" checkpoints because their power rests in providing you with the ability to sense, or "*feel*" the effectiveness of your other checks all at once without having to consciously check each individually. Second level checkpoints don't replace first level checkpoints, rather they provide feedback on how effective the other checks you may be doing are working. You can't get this information from first level checks alone. They provide different and valuable information.

> **Framework Vision: Examples of challenges where the framework can help**
>
> Examples of challenges the framework can help you with in business include ensuring you have addressed all the key technical risks, resolving conflicting requirements, and ensuring you have agreement of key stakeholders before moving forward with critical decisions. These types of challenges relate more directly to your performance.

First level checkpoints are routine. They don't require a great deal of time or energy to implement. For example, I check my grip, then check my club face, then check my shoulders and stance. In business in organizations that use an agile approach this is analogous to holding a short daily standup meeting at the start of each day. Each of these checks takes only a short period of time, but they help ensure we are prepared for what we are about to do in the near future. Second level checkpoints, on the other hand, take more time to learn and execute because the actions you might take are not always straightforward. There are usually options and consequences that need to be considered with second level checkpoints.

# 2nd Level Example: Requirements Incomplete or Ambiguous

A common business example of a second level checkpoint is recognizing your requirements are incomplete and/or ambiguous. Often the best way to sense this is by listening to the informal discussions of the people on your team. How you handle this type of situation is not always straightforward because it depends on many factors.

Do your key stakeholders know the requirements are incomplete and/or ambiguous and are they willing to work with your team to evolve a sound set of requirements? If so, the best option may be to take no action. But if you have a firm fixed price and schedule, and non-collaborative stakeholders, then the consequences of inaction could be deadly.

You may need to dig deeper, asking more questions before you decide on the appropriate course. Have you worked with these stakeholders before? Can you talk to them and explain the situation and come to an agreeable approach to solve this problem?

The way you get better at handling these types of difficult situations is partly through experience. However, you can also get better by sensing patterns you have seen before, and recalling the options that have worked best in the past. But doing this takes practice.

## Summarizing the Essentials of Second Level Checkpoints

Second level checkpoints:

- Keep our focus on the goal by providing an objective external cross-check.
- Always include rapid feedback and refinement of resolution ensuring your changes are moving your performance in the right direction.
- Require more than a yes/no answer. They require practice because there are always options to be considered and a decision to be made.
- Can be made easier by creating a small set of patterns and learning to keep the patterns fresh in your mind (more discussion in next few chapters).

# Making Second Level Checkpoints Easier Through Patterns

The small set of patterns required to help your second level checkpoints can be captured in different ways:

- If you have a good memory, you can keep them tacitly in your head.
- You can use a table that includes reminders of actions to take when each pattern is observed, as we saw in the procurement example earlier in this chapter.

> **Framework Vision: Helping you recall your own common situations**
>
> We envision that you will be able to add your own common situations to the framework to aid your practitioners in their decision-making.

In the next chapter we continue our investigation into second level checkpoints helping you understand why they are essential to achieving and sustaining higher performance.

## Chapter Seven Summary Key Points

- "If we can't identify what decisions would be affected by a proposed measurement and how that measurement could change them, then the measurement simply has no value."
- Don't ask your people to collect data. Empower the team to find and solve specific performance problems.
- Second level checkpoints:
    - Keep our focus on results by providing an objective external cross-check thereby ensuring our first level checkpoints are meeting their intent.
    - Always include rapid feedback and refinement of resolution ensuring your changes are moving your performance in the right direction.
    - Require more than a yes/no answer. They require practice because there are always options to be considered and a decision to be made.
    - Can be made easier by creating a small set of patterns and learning to keep the patterns fresh in your mind.

# Chapter Eight: Feeling Your Way to Higher Performance

*"I don't know how I know where to pass. There are no firm rules. You just 'feel' like you are going to the right place….And that is where I throw it." Tom Brady, Quarterback, New England Patriots*

In this chapter we continue our investigation into second level checkpoints helping you understand why they are essential to achieving and sustaining higher performance. Specifically, you will learn how second level checkpoints are more of a *preventative aid*, as well as an objective cross-check, and you have to get to prevention if you want to achieve and sustain higher performance. [7] You will also learn a practical technique that can help you keep your eye on your goal.

How the second level feel checkpoints work with my golf swing as an "objective" measure requires some explanation. In my case there are actually two second level checks that I use. We have already discussed the check I use right after the golf shot to take action based on the result I observe. This is similar to the procurement case where I am ready to rapidly sense a small set of likely patterns following each shot (shot going short, shot fading), and I know the possible actions I can take that have led to positive results in the past. There is another second level check I do just before each shot.

## You Have to Get to Prevention for Higher Performance

The "*feel*" check before the shot senses multiple triggers at once. This is similar to the business case previously described where keeping the big picture feel of the project plan continually in one's mind could have helped my client avoid doing unplanned work and failing to do critical planned work. As I visualize the planned shot different parts of my body let me know if they are each prepared to execute. In business this is comparable to continually listening to your team. If I don't get the proper feedback from key body parts to my brain it is an immediate signal something is off. The best project leaders sense discomfort among the team and learn to take rapid action before serious damage occurs. Knowing how to look for and sense early warning signals takes us beyond simple first level checks. Second level checks help us avoid problems before they occur which is far more effective than dealing with their impacts afterwards.

> **Sidebar: Lean Process Analysis**
>
> When conducting a Lean Process Analysis (technique found in the Lean Six Sigma Toolkit) all non-value-added activities are highlighted. This includes any activity associated with detecting errors after they are made because the goal is zero defects, or avoiding problems in the first place. Sometimes we need to put checks in place, but the first goal for high performance is learning to not make the mistake the first time.[34, 35]

On the surface one would think that the "*feel*" of a golf swing, or the feel of a project plan, is too subjective. But the reason this works is because we are all susceptible to falling into poor habits when relying on first level checks alone. In my case I often find myself thinking I am placing my thumb and club head in the correct position, but not really checking it correctly. In business this is like the project leader asking if a task is done, rather than taking the time to listen to the concerns of the team. In this sense second level checks are more *objective* even though they may on the surface appear more subjective. They are more objective because they are based on a different perspective providing a degree of cross-check.

# Sensing the Wrong "Can Do" Vision for High Value Payback

Furthermore the second level checkpoints are more effective at preventing the kind of trouble that hinders organizations from achieving the high value payback they seek. As a quick example I recently participated on a software review team for a major US defense contract. This project had a long history of missing schedule commitments and cost overruns. When we asked the people how they estimated the work remaining, it was clear their assumptions were based on a perfect world where nothing else would go wrong despite the fact that the history of this project indicated no such pattern in the past. It is great to create a "can do" vision in your organization to motivate your team to do the best possible job. But it also shows lack of sound management and leadership when you ignore years of clear historical objective data when projecting future performance. Observing such patterns early when you can take action to prevent serious future trouble before you are forced to take action because you have no alternative, is the power of second level checks.

Some may think that second level feel checkpoints are soft. In fact when I first started to discover the power of second level feel checkpoints even though I knew they were working for me I was worried that I would not be able to communicate their value to others. This was because the idea of checkpoints that didn't have clear simple yes/no answers seemed counter to what I had previously been taught. The second level checkpoints are more complex because they tie to situations where decisions must be made with multiple possible options. At the same time I realized that my experience in business over many years from helping many clients indicated that the use of the simple first level checks alone had failed over and over at helping organizations move beyond a fundamental level of performance. I had actually thought I was discovering something new with the idea of the more advanced second level "feel" checkpoints until I picked up a book by Jonah Lehrer.

## An Example of "Feel" Leading To Better Decisions

Jonah Lehrer in his book "*How We Decide*" [20] tells a story about Tom Brady, the quarterback of the New England Patriots in the National Football League. In the 2002 Super Bowl, the Saint Louis RAMS were heavily favored. There was one minute and twenty-one seconds left on the clock and the score was tied. The ball was on the Patriots 17 yard line. Tom Brady, the second string quarterback at the time was brought into the game. He was the 199th pick in the 2000 draft. The scouting reports didn't expect much of him saying he was skinny with a poor build. There were only a few words mentioning Brady's positive attribute: "*decision-making*."

Quarterbacks have to make quick decisions on the football field. They have to detect quickly the weak spots in the defense as each play unfolds. They need to assimilate *approximately* where every player on the field is. As the last minute and twenty-one seconds unfolds Brady makes the right decisions, one after the next. During the first few plays he completes short passes just taking what the defense will give him, but as he does he marches the Patriots closer and closer to field goal range and eventually close enough for Adam Vinatieri to kick the winning field goal as time expires. When asked about his decision-making ability Brady responds:

*"I don't know how I know where to pass. There are no firm rules. You just 'feel' like you are going to the right place....And that is where I throw it."*

## Feel Versus First Level Checkpoints

The problem that I saw with that client who had trouble planning is similar to what I often see when people first decide they want to get better at golf. They fill their heads with 100 things to think about thinking they are all important to hitting a golf ball. This is because they read about those 100 ideas in books. But the problem is that while each of those 100 things may have some value, they don't know how those 100 things tie to their specific goal of getting better under their specific conditions under which they must operate. And as a result it doesn't work. This is a primary weakness of first level checks. They create extra effort that often is not directly related to the specific goal.

> ## Sidebar: Business planning reminder
> When planning make sure the plan for your measures *feels right*, you can commit to those measures, and you can clearly see how the measures align with your objectives and can help your team perform. Keep in mind that you may only need a very small number of measures to achieve your objectives. Refer to "How To Measure Anything" by Douglas Hubbard.

This would be like Tom Brady as he fades back to pass trying to explicitly look at where each of the other 21 players on the field currently are to use that information to calculate where he should throw the football. The problem is he doesn't have time to do this. He doesn't have time to calculate all the possibilities. He has to figure out a way to feel the right answer and have confidence in what he feels.

So are we saying that Tom Brady should just ignore where all the other players are and just listen to his emotional brain telling him where to throw the football? Are we just hoping he guesses right? Of course not. Tom Brady went on to perform as he did in that 2002 Super Bowl again and again. It isn't just guesswork, or luck.

> **Glancing Forward**
>
> Some may think this is leading to "just trust your gut feel". This is not at all the case being made.

# Practice to Help Keep Your Eye on Your Goal

Now let's go back to that client discussed earlier in the book that needed help connecting the theory of planning to his real project conditions and show you how we helped him to better prepare himself for higher performance.

In the planning scenario I now use in my workshops, a new project leader is told he must get work done rapidly because he has a very aggressive schedule, and the customer will be in plant in a couple of days. He also learns that the people who were planned to be available at project startup aren't really available. He is then asked to explain what he is going to do as the class listens. We then walk through a simplified five step planning process and discuss how each step can be accomplished even if you only have a small amount of time. We also discuss the goal of each step.

By connecting each step to its goal and to the real world scenario we help each participant see how to connect the theory of planning to the conditions they are likely to face when they leave the workshop and go back to their project. The scenario discussions in the workshop lead the group through *"how to"* specific options within the planning activity, and group brainstorming almost always leads to some interesting new options and consequences to consider.

> **Framework Vision: Helping your team with key competencies**
>
> All successful teams need key competencies including how to plan and scale a plan. Our framework should identify key competencies such as Stakeholder Representation, Analysis, Development, Testing, Leadership and Management. Planning and scaling a plan should be included as part of the Management competency.

A technique I suggest is to write down on a sheet of paper 3 to 7 of the most important pieces of information related to the project's (or your current iteration's) vision and the plan to achieve the vision and tack it up on your office wall. I then suggest that this be used as a reminder so the participants can quickly refer to it on stressful busy days when faced with decisions that must be made in less time than they would like. (including decisions related to which task is most important right now).

> **Framework Vision: Helping your team with priority decisions**
>
> Our framework will not magically give you answers to all your challenges, but it will provide a practical and simple way to rapidly remind people what is most important to be focusing on right now, and it will provide a structure under which you can add more specific information to help you find your own answers to your challenges. The rationale for this framework need is based on the observation that too often practitioners are faced with too much work on their plate and they often need help in deciding where the priority should be placed.

One insight we have discovered through these workshops is that first level checkpoints work well during preparation and planning steps, but they tend to obstruct progress when used during execution. When I was trying to explain second level checkpoints to my daughter (a flutist), she said as a musician it was best not to consciously think about details when you actually played your instrument. Her words were that it was better to try to get into the zone, or the essence of the music. She said you need to trick yourself into feeling like you are letting go of the details, but you don't really. If you try to hold onto them, they will escape you.

## Fundamental Eleven:

Just following a process isn't enough to sustain high performance. It must be the right process that addresses the real goal.

Today through breakthroughs in neuroscience we understand more about how Tom Brady does what he does, even though he says he doesn't understand it himself. When Tom Brady fades back to pass rarely does the play come off exactly as diagrammed. How Tom Brady makes these rapid decisions under pressure and constantly changing conditions, why he does it better than most, and how you can learn to do it too in both your professional and personal performance endeavors is the subject of the next chapter.

## Chapter Eight Summary Key Points

- Second level checkpoints are more objective because they are based on a different perspective providing a degree of cross-check.
- Second level checks help us avoid problems before they occur which is far more effective than dealing with their impacts afterwards.
- Just following a process isn't enough to sustain high performance. It must be the right process that addresses the real goal.

# Chapter Nine—Better Decisions Through Better Practice With Patterns

*"We cannot solve our problems with the same kind of thinking that created them"*
*Albert Einstein*

In this chapter we continue our investigation into a technique I refer to as *integrated practice* and I explain how this different form of practice can help you make better decisions leading to higher sustainable performance.

Part of achieving and sustaining high performance is making better decisions. So how does Tom Brady make rapid decisions under pressure and how does he make them better than most in complex situations? Lehrer tells us that our dopamine neurons, or the emotional part of our brain, love patterns—and they are key to helping Tom Brady make rapid decisions. But Tom will only make a good decision if he has the right pattern fresh in his brain at just the right time when a certain play is run against a specific team.

How does Tom Brady ensure the right patterns are fresh in his brain when he needs them most? He spends hours looking at game tapes before the next game. And not just any game tape. Tapes of the specific team he is going to face next, and tapes of the most recent games they have played both against other teams and his own team. He is looking for very specific information as to the routes his opposing team members frequently take, and plays they run when faced with situations similar to what they will face against his team. He isn't just looking at fundamentals, or general statistics. He is looking for their specific strengths that he wants to avoid, and he is looking for their current specific weaknesses that he wants to exploit. This is a specific kind of information, and he knows it is precisely the information he needs to have fresh in his mind at game time so he can rely on his instincts under pressure to help him make the best decision. As we move forward in the book this *specific* kind of information is also referred to as *contextual* data.

## What Is Integrated Practice?

So just what is *integrated practice*? First, it is a form of practice that is different from what many of us have been taught about practice in that it is integrated much more closely from a time perspective with actual performance. Second, it is different from what many of us have been taught in that a significant part of integrated practice is *mental* practice or simulation of our planned performance in our heads. We are preparing mentally for what we are about to do in the very near future. Why this practice needs to be close in time to the actual performance may already be obvious to you, but I will explain it further in this chapter and the following chapter as we examine more closely key attributes of integrated practice including objective and contextual data. Both are essential to effective integrated practice leading to higher sustained performance. Let's start with a discussion of patterns.

## What Is a Pattern?

A pattern– in the general sense as I use the term throughout this book– is an abstraction of a common situation that occurs in a specific context (could be good or bad) with one or more possible alternative actions.

In his book, "High Performance Operations" [37], Hillel Glazer refers to patterns as "structures and guidelines for how work is to be done." Hillel also tells us that "patterns have typically replaced what most operations would have previously called processes. While policies most appropriately convey values, patterns may also convey values, but are more aligned to principles." [21] [38]

## The Good And The Bad Side of Patterns

To help understand the importance of integrated practice we need to discuss more what neuroscience has helped us understand about the way our brain works. [20] The emotional side of our brain—specifically our dopamine neurons– help us by helping us recall patterns rapidly. But if we train our dopamine neurons with the wrong

---
[21]Patterns are also related to the popular notion of "thin slicing".

pattern they will lead us exactly where we don't want to go. This is why people sometimes make bad decisions when they have to respond quickly. If the tapes we've been playing in our head—especially most recently– are not the right patterns it is likely to lead to the wrong decision on game day.

An example that Lehrer points out is the inaccurate pattern many of us have in our head with respect to credit cards and the stock market. Too many of us believe when the market is going up it will just continue to go up, and too many of us believe that when you hand someone a credit card it isn't the same as handing them real money. These false patterns we put in our heads lead many to make poor decisions leading to stock market bubbles and personal credit failures.

# Creating the Right Patterns: Software Developer Examples

Often when we are trying to improve our performance we find we can do something well under one set of conditions, but when the conditions change our ability to perform degrades. An example from the business world is the software developer who is great at testing when given adequate time to methodically conduct all planned tests, but needs more help when faced with too little available schedule time. Another example is the software developer that makes few programming errors when the requirements are clear and stable, but needs help when faced with conflicting or incomplete requirements. To get better and keep getting better we must learn to perform under the varying conditions that we will ultimately need to face some of which often turn out to be adverse[22] [39].

---

[22]For more information about varying conditions refer to Myburgh's Situational Process Model (SPM) evolved from the original Situational Software Engineering model described by Myburgh in 1992.

> **Framework Vision: Critical thinking**
>
> Our framework will help practitioners recall common situations they should be alert to, and possible options and consequences to potential related decisions. In this way it will become an aid to critical thinking.

We have discussed in the previous chapter the value of taking simple checks to a higher level and how this can help us sense when conditions may be changing by practicing the use of a big picture feel. But most people need help creating the right patterns leading to the right big picture feel that can help us rapidly sense when corrections are needed.

# Recalling the Procurement Case

Recall in the procurement case discussed earlier in the book we created five scenarios for the procurement specialist to keep in the front of his, or her, mind helping to quickly recognize the appropriate action when each scenario occurred. This technique can help create the right patterns[23] leading to better decisions.

Another technique that can help as you keep improving is to continue taking notes as this will help you recognize new patterns or variants of old patterns specific to your own performance and environment. These new patterns can provide on-going opportunities for continued performance improvement. This technique helps you create the right response to the most likely scenarios you will face as long as you keep those patterns fresh in the front of your brain when you need them.

By refining your patterns and learning how to keep the right patterns fresh in the front of your mind you are making the tapes you go over in practice more accurate to the conditions you will face on the job which in turn will help you make better decisions like Tom Brady. The way you keep the right patterns fresh in the front of your mind when you need them is to practice at the right time which is close in time

---

[23] As stated previously in this chapter, a pattern is an abstraction of a common situation (or scenario). In this case we captured the patterns in a table format. How you capture the pattern is not what is important. Rather, what is important is that you learn to recognize the pattern when it occurs and you recall your options and consequences related to possible decisions.

to when you perform [24] whether that be a musical instrument, a sport, or conducting an activity on the job.

> ### Sidebar: General
>
> As stated in the introduction to this book, "this is a book that can show you how to get just a little bit better at whatever you want to get better at and keep on getting a little bit better each day." The sports analogies in the book are only used to emphasize the criticality of continual practice to improving and sustaining improvement. Analogies with world-class athletes or musicians should not be taken further as comparisons in other areas may break down.

# Spreading Positive Performance Across Your Organization

Causal Analysis and Resolution (CAR) is a CMMI Maturity Level 5 Process Area. Its purpose is to identify causes of selected outcomes and take action to improve process performance. A tip in the CMMI Guidelines book tells us that *"CAR helps you establish a disciplined approach to analyzing the causes of outcomes, both positive and negative, of your processes."* Historically many organizations have focused primarily on problems, or negative outcomes, when applying CAR. But if you want to see performance improvement one of the best approaches is to use CAR to raise up the visibility of scenarios that have been proven to lead to higher performance and spread them across your organization. A simple technique that can help is capturing positive patterns and using them as quick reminders when faced with similar situations that require on-the-job decisions. By placing more emphasis through CAR on scenarios that enhance positive outcomes we increase our chances for sustaining high performance.

---

[24] With enough time inserted between the practice and the performance to ensure you are adequately rested.

> **Glancing Forward**
>
> In Part III of this book we provide an example framework along with common scenarios that require practitioners to make on-the-job decisions that potentially could lead to negative outcomes due to poor decisions. We also identify options that could be taken leading to positive outcomes. There are additional scenarios found in the appendices to this book including how to handle ambiguous or conflicting requirements, what you can do when you can't get all the data you need from an external source, and how to handle a risk that isn't being properly dealt with by the team.

Let me now give an example that demonstrates a technique you can use together with contextual data that can help you develop your own practice patterns leading to more positive outcomes.

# Utilizing the "Listen-for" Technique For Positive Outcomes

One of the most common questions I have received in the past from project leaders is "*how do I get accurate status from my people?*". And they have often followed up that statement with something like, "*...without bothering them?*" Today, with the increased use of self-directed teams more team members are asking this question because accurate status is something all team members need to know.

The difficulty faced by many leaders and team members is how to go about ensuring they have accurate status, but doing so without negatively impacting the on-going work. Having accurate current status is important in the business world to organizational leaders, but it is also important for individuals in assessing where they are in their own work tasks. Accurate current status helps us all know when its time to make a correction. I discovered the technique I am about to describe when I asked a manager who was known for being a very effective leader in one of my client organizations:

*"How do you get accurate status from your people?"*

He replied,

"*When I go to talk to my people I listen for...*"

And then he went on to list a number of specific things he knows to listen for. As I listened to him describe this process he uses, it occurred to me that what he was telling me was that before he stops by to talk to one of his workers, even though he makes it appear as just a casual unplanned visit, he has actually prepared ahead of time by thinking about the most likely issues that could be occurring on this specific project right now that this worker is involved with. I have referred to this technique in previous publications as the "*listen for*" technique. [31, 32, 40] This technique can be used to help team members prepare for daily standup meetings, and retrospectives. In this book I refer to this specific data also as contextual data.

Note how this is analogous to Tom Brady reviewing films of the team he is going to face next just before the game. Tom is looking for specific data that relates directly to his immediate context which he refreshes in his mind so the emotional side of his brain has the most accurate patterns available just when he will need them to make a quick decision.

For business people, when you are preparing for discussions with colleagues, managers, employees, customers, subcontractors, or other stakeholders, think about and write down three or four of the most common scenarios that occur during this specific phase of the project related to the specific type of activity this individual is currently involved with. Ask yourself: *Are there certain scenarios that are more likely because of any specific conditions occurring at this time?* When doing so, refresh in your head any past similar project or personnel scenarios you have recently experienced. Think about likely consequences you may want to bring up in the discussion.

## Sidebar: How much information is required for each pattern?

The amount of information you need to capture for each pattern depends on your specific situation and is therefore left up to the team to decide. With some teams this information can be kept tacit, while other teams will decide there is value in capturing more written information.

In workshops I ask participants to do this as an exercise. Then I have them jot down strategies that have been used in the past that have been most beneficial at resolving the most likely issues that will arise. Then before you initiate a conversation review this material so it is fresh in your mind, but I suggest it is best not to ask specific questions associated with the data. This helps to keep the conversation open and free-flowing.

## Example Using Patterns to Aid Personal Improvement

For personal efforts, such as improving your personal performance at your favorite sport, think about the typical scenarios where you most often get into trouble. If you are a golfer, are there specific holes that you always have trouble with? Do you typically play well on the first nine holes, but then fall apart on the second nine? Do you struggle when your competition talks too much during the round? Then start thinking about what you might do differently when these specific conditions occur. By putting more specific and relevant patterns in your brain just when you need them you improve the likelihood that you will make better decisions leading to higher performance.

## Example Using Patterns To Improve Business Performance

Similarly, in business, if there are specific situations that reoccur and you frequently have trouble with, create the common scenarios, and then think about how you might behave differently to get a more positive result in the future. Then start practicing by going over the scenario in your mind close in time to when the situation is likely to occur.

Most people know what to listen for, and we know the typical scenarios that have hurt our performance in the past. As an example, keeping this information fresh in our minds can help to uncover key information that might never come out in a typical project status meeting, and in turn can help us make better decisions leading to improved performance.

> **Sidebar: Being responsible doesn't imply competency to perform**
>
> There may be common scenarios you have observed with your teammates on your agile team. With self-directed teams each team member is responsible to raise issues, concerns and risks. But being responsible doesn't imply competency to perform.

> **Framework Vision: Helping teams rapidly evolve their competency**
>
> This is an area where using our framework could help you and your teammates rapidly evolve your competency by recognizing situations faster, along with their options and possible consequences, leading to better and more timely decisions.

Keep in mind that it is best to think about weaknesses and first level checks ahead of time, but forget them when you actually perform. Consciously working on known repeating weaknesses is best done through the rational side of the brain when you are preparing for an activity. Then forget them and trust that you have properly prepared your emotional brain for your best performance. A mistake many people make when communicating with others, especially those who they may be giving direction to, is to use too much of their rational brain at the wrong time asking pointed questions that can put a worker into a defensive posture shutting down communication when they seek to open it up.

These techniques may seem obvious to you, but not everyone is good at executing them and I have found that many who agree they make sense, often just don't take the time to use these practice aids. It is also important to understand that because the environment we must perform in is constantly changing (phase of the project, current issues, risks) you must continually refresh the patterns with the latest data. This is partially why we stress that this practice must be integrated—that is conducted close in time—with actual performance.

> ### Framework Vision: Keeping your practice aids current and useful
>
> The framework must be easy to access, use and update as practitioners learn new things interacting with their teammates each day on the job. The rationale for this framework need is based on the observation that if it isn't easy to access, use and update, it simply won't be used by busy practitioners.

## 🔑 Fundamental Twelve:

Even if we know it makes sense to practice, it won't help if we don't discipline ourselves to do it consistently at the right time.

If Tom Brady fails to study tapes of his next opponent during the week leading up to the next game he will not be as prepared as he could have been and will be at increased risk of making the wrong decision that could cost his team the game. Recall fundamental One and why we distinguish integrated practice from training. Training helps you understand expectations related to a job. Integrated practice helps you learn to actually do that job in the environment in which you must perform.

This is Tom Brady watching what the team he is facing next did last week, and watching it just before the game. He can't stop looking at those films for even one week or his performance will degrade. He has to keep looking at the most current films of his next opponent because the environment keeps changing, the players keep changing, some get hurt and others get traded, the plays his team runs and the plays the opposing team runs keep changing. The more specific your data is to your context and the more current your data, the better the chances of making the best decisions on game day.

# What Does Practice Have To Do With Performance?

When the situation changes, the objectives, and measures may need to change. Project conditions keep changing. Team members change, customers change, stakeholders change. Keep asking yourself:

- What are the critical objectives today?
- Have the critical areas we should be focusing on changed?
- Are we looking at the right information?
- Is the information we are using relevant and specific to your current situation?

If you aren't continually practicing with current data based on current situations, don't expect to sustain high performance.

Using contextual data can help you create the right patterns leading to more effective decisions and higher performance, but you can also be fooled by your data into believing you see a pattern that isn't really there or believing the pattern you see demonstrates a root cause when in actuality it is only a symptom. We can be fooled because the emotional brain wants to see patterns so badly it will even make one up if you haven't refreshed your brain very recently with current accurate data. What we can do to ensure we are not deluding ourselves with false patterns leading to poor decisions is the subject of the next chapter.

## Chapter Nine Summary Key Points

- Part of achieving and sustaining high performance is making better decisions. We will only make better decisions if we have the right patterns fresh in our mind at the right time.
- Integrated practice is "mental" practice or simulation of our planned performance in our heads with the right patterns based on the right contextual data, executed at the right time.
- Often when we are trying to improve our performance we find we can do something well under one set of conditions, but when the conditions change our ability to perform degrades.
- Even if we know it makes sense to practice, it won't help if we don't discipline ourselves to do it consistently at the right time.
- When the situation changes, objectives, & measures may need to change. Project conditions change. Team members change. Customers change. Stakeholders change.
- If you want to sustain higher performance, you have to keep practicing, and you have to do it close in time to actual performance with the most current contextual data.

# Chapter Ten—Right Patterns Through Practical High Maturity

*"No data have meaning apart from their context."* Don Wheeler

In this chapter I explain how you can use the CMMI high maturity practices and a technique I refer to as a *"Structured Real Story"* to help ensure you have the right patterns leading to sustainable high performance.

The CMMI high maturity practices (level 4 and 5 practices) are intended to help organizations improve their performance. Nevertheless, many organizations have struggled to implement these practices in a way that has proven to provide real value to both their organization and their practitioners who must perform each day on the job.

While the techniques shared in this chapter may be viewed as non-conventional, they meet the intent of many of the CMMI high maturity practices and they also have been proven to provide high value performance benefits to both organizations and their practitioners. But before discussing these techniques I want to motivate why they are needed today by reviewing a few key points already discussed in this book and providing a few more examples that demonstrate the significance of the problem we are facing.

## Motivation for the Techniques Discussed in this Chapter

We have previously discussed how you can achieve better decisions and higher performance through the use of contextual data and integrated practice. But if the contextual data you collect is not accurate it will not lead to better decisions. To ensure your contextual data is also accurate we need to ensure that at least some of it comes from an objective source.

Tom Brady succeeds because his patterns are based on game tapes of his opponent's recent games and his own team's recent games. Those tapes provide objective information because they are coming from an external and reliable source[25] —and it is the right information he needs to perform on game day.

On my personal improvement project, over time, my first level checks degraded at least in part because I was relying on my own memory to execute them correctly. I was relying on words in my notes I thought I understood, but over time I found myself misinterpreting these words.

Furthermore, as discussed earlier in the book I was still unsure as to whether I had accurate information with respect to the real root cause of my repeating pain points. I knew I needed to keep working to try to uncover the real root cause, otherwise those pesky repeating weaknesses would just keep coming back and hurting my performance at the worst possible time. Let's now look at what I was doing as I continued to search for the real root cause of my pain and lets also look at the affect this process was having on my performance.

## In Search of the Real Root Cause on the Personal Improvement Project

In my 2009 notes I found myself questioning the meaning of certain words. For example, I started asking myself, what do I mean by "Square to the target?" Sometimes I thought I was "square to the target" with my shoulders when actually I wasn't. The same problem would happen with the club head. What did I mean when I wrote "keep the club head square?" I would think it was still square when in actuality over time I would have started laying it down facing to the right of the target just repeating my old poor habits again.

---

[25] Objective data can also be found by looking at the same data from a different perspective. Refer to earlier discussion on Second Level Checkpoints.

> # Framework Vision: Terminology
> 
> Our framework will provide a common terminology that can help your organization communicate progress and health more consistently and accurately. Some of the terms chosen may be questioned because the framework will not always use terms commonly used today. The reason for this is because conscious decisions are made not to use certain terms that may communicate an inaccurate meaning.

As a result of these small errors my shoulders would gradually open up and I could not make a complete backswing. This, in turn, would lead to my swing getting shorter. When this would happen, because it was so gradual, and because I had no external objective reference, I wouldn't consciously notice it at first. But subconsciously my body would sense that something was wrong. Something didn't feel right. As a result parts of my body would try to help make things better. For example, my hands and arms would start moving faster. But my hands and arms couldn't fix this problem because they didn't create the problem, and their attempt to fix it only made matters worse. My hands and arms getting in front of my body would lead to pulled shots to the left or sliced shots to the right. When all this happened the parts of my body all felt like they were fighting each other. Refer to Figure 10-1.

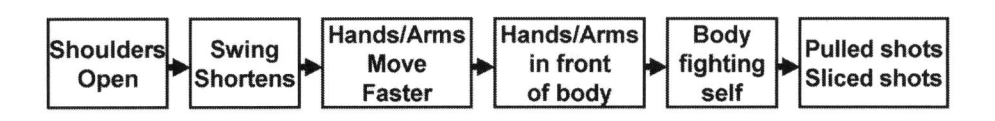

Shoulders Open → Swing Shortens → Hands/Arms Move Faster → Hands/Arms in front of body → Body fighting self → Pulled shots Sliced shots

Figure 10-1 Chain of Events: Small Errot at Start Leads to Amplified Impact

# The Organizational Performance Impact of Inaccurate Data

Have you ever noticed a similar pattern in organizations? A small problem is misdiagnosed because we don't have accurate data leading to an amplified chain of events that only adds to poor organizational performance. If we could just learn to catch more of these situations at the right time and then take the time to ensure we had accurate data leading to the best decision at that point in time, the cost savings could be enormous!

This may sound simple. It isn't. The reason is because the right process to address root causes of repeating specific weaknesses is not well understood in most organizations. It must start with ensuring we have current contextual and accurate data.

# Business Example: Impact of Taking Action with Inaccurate Data

A common scenario in business of a small problem misdiagnosed because we don't have accurate data leading to an amplified chain of events that only adds to poor organizational performance is the dictated schedule. It usually goes something like this:

*Project Leader speaking to her team: "Management has dictated that we will not slip schedule, so everyone needs to do whatever it takes to get the job done."*

*Developer response: "OK. I am just going to get my code to run. I am going to skip my design review and I'll skip most of the testing I was going to do because management doesn't care if I follow the process."*

This scenario I have seen play out multiple times, and when I have investigated these cases by asking a few direct questions (looking for accurate data) I have often found that management did not intend their message to imply that they don't care about following the process.

Developers need more than just the technical competencies such as analysis, development, and testing. They also need an appropriate level of management competencies to make decisions related to self-managing their work. This will help them know where their options lie, and the consequences of their decisions. They also need to keep scenarios such as this one fresh in the front of their minds especially during times of high project stress. This scenario most often leads to longer schedules, rather than achieved schedules. Refer to Figure 10-2.

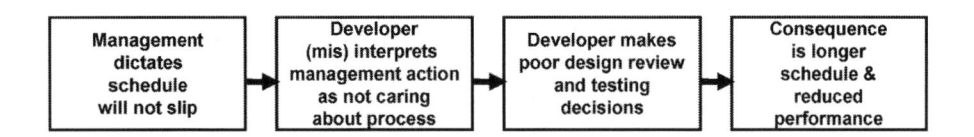

| Management dictates schedule will not slip | → | Developer (mis) interprets management action as not caring about process | → | Developer makes poor design review and testing decisions | → | Consequence is longer schedule & reduced performance |

Figure 10-2 Performance Impact of Inaccurate Data: Dictated Schedule Scenario

So how do we avoid these kind of common scenarios that repeatedly hurt our performance? If you are going to use your data to drive your decisions you need to do everything possible to ensure your data is as accurate as it can be. We now know- when it comes to repeating weaknesses- the data we collect may never be perfect. So what can you do to make it as accurate as possible?

# The Power of Small Changes

You probably know the answer by now, but it bears repeating. You may not know the exact root of your repeating weaknesses, but continual small changes based on rapid *objective* feedback has been proven to be the most practical way to sustainable high performance.

My experience in business indicates small changes are often the most valuable changes when it comes to performance. In fact, a small change, if instituted at just the right time, can not only sustain performance, but it can improve it, and if you keep making those small changes you can keep on improving it. We saw an example of this earlier in the book with my client that focused on just a small set of measures, but ensured the measures had objective backup, and moved their process improvement focus to the "little things" that could help their people on the job. By continually staying focused on even a small set of changes high value improvements will happen and they can keep happening.

On the personal improvement project, the good part was that because I was continuing to take notes and analyze them, and make small changes, I now had a better understanding of the cause of my recurring pattern. An example of objective data that helped my performance was observing the results of each shot, such as the golf ball fading, or going shorter. These became signals that I could sense quickly right after each golf shot, and I could make a rapid adjustment to keep the problem under a level of control.

> **Sidebar: The definition of insanity**
>
> *"The definition of insanity is doing the same thing over and over and expecting different results."* Ben Franklin. I include this quote as a simple reminder that making small changes is not the same as doing the same thing over and over.

## Reinforcing An Important Insight About Small Changes

Note that I actually had achieved my performance objectives in 2008, but I could not sustain that performance in 2009. I had not yet discovered why this was the case. This happens with organizational improvement efforts as well. This leads to the next Fundamental. If it is not already clear, in Chapter Twelve when you learn what I discovered about the root cause of my personal weaknesses you will understand why this Fundamental is so important.

### Fundamental Thirteen:

Once you have achieved your performance objectives you have to keep making small changes, and you need a mechanism in place to rapidly sense the effects of those small changes, and rapidly respond to those effects to minimize performance impact.

> ## Framework Vision: Practitioners owning their practices
>
> Our framework approach supports fundamental thirteen by placing practitioners in control of their own practices and giving them a mechanism to keep their practices current with the information they need to continue to maintain high performance.

# Example of Small Change To Counter the Dictated Schedule Scenario

Fundamental Thirteen is not intended to imply just change anything. The changes you need to be making must be specific to address the continually changing world around us. An example of a small change that could be made in the Dictated Schedule Scenario is for the developer to assess the work remaining to complete the planned design review and testing. Then assess possible options, such as focusing the design review on high risk areas only in order to reduce the overall design review effort and schedule, and then reassess any related risks before making this change.

# Helping Your People Make Effective and Timely Small Changes

Whether or not the approach described in the previous paragraph is actually a change to your process depends on how the processes are written in your organization. When I help clients develop processes in organizations that want to encourage increased agility and process ownership by practitioners I encourage them to write their processes in a way that supports increased practitioner decision-making.

As an example, one way to do this is to provide *criteria* in the process to guide practitioner decision-making.[23] Criteria that could be used to help practitioners decide when a peer review is required could be:

- complexity of change
- experience of developer
- history of component

Using such criteria gives the practitioner options based on their specific context.[26] Providing examples of common scenarios with options and consequences of likely decisions can also help because it reminds the practitioner of the alternatives available to them and it motivates them to think-through the possible consequences of each. These types of process aids have proven to be a practical way to empower practitioners to make the small changes that are needed for sustained performance without a need for formal process deviation approvals up the organizational chain.

Scott Ambler told me that in his experience many teams have no idea what small changes they could make, and that they often struggle to identify root causes of their problems because they don't understand the tradeoffs they are making when it comes to their practices. After Scott's Disciplined Agile Delivery book [22] came out, he introduced "goal diagrams [41] to help people understand their options and the tradeoffs involved. This in turn helps them identify potential small changes they could make.

Next I want to talk about the common measurement mistake discussed earlier in this book in Chapter Three. Then I want to talk about what you can do to avoid this problem. This will lead into a discussion of a technique that can help ensure once you have the right measures that you use them in a practical way to help your performance.

## Revisit the Common Measurement Mistake from Chapter Three

Earlier in the book we heard a data analysis scenario that went like this:

*Our data indicates we are making too many mistakes during the requirements and design phases. The data indicates we should have been finding these defects earlier*

---

[26] Another approach can be found in the Disciplined Agile Delivery (DAD) framework. DAD has a goal-based approach that employs goal diagrams. More importantly it contains tables summarizing the tradeoffs of each technique called out by the goal diagram. This helps people understand their options and the tradeoffs involved. This in turn helps them to identify potential small changes they could make.

*during a peer review so we need to improve our peer review process of our designs and requirements. Our people also need better training so we should update our requirements and design training with the latest new techniques.*

And we heard a common consequence from a practitioner was something like:

*I got told I had to attend some new training course that was suppose to help, but it was basic stuff that I learned years ago and it had little to do with what I face each day on the job.*

What might occur if an analyst *followed this thread* a little deeper asking a few more questions?

**Analyst**: *So I see there are many requirements defects, but it also appears that going to requirements training didn't help. Why do you think this training didn't help you?*

**Requirements Engineer**: *Because the problem isn't that I don't know how to analyze requirements data. The problems we have with requirements I am dealing with right now is that we can't get the data we need from the customer. He wants us to just do the best we can, and then when he sees what we developed he will write up defects to tell us what he really wants.*

**Analyst**: *Sounds like you have a risk. Have you raised the risk to your management?*

**Requirements Engineer**: *Raising this risk to management doesn't do any good because they don't know how to solve this problem.*

**Analyst**: *So are all your requirements problems caused by data problems?*

**Requirements Engineer**: *No. There are other issues we could solve if we could get the customer to help us clarify a number of ambiguous requirements. But when we try to get them to commit, they won't work with us so it leaves us open to more late defects.*

**Analyst**: *Sounds like you've been through this before and you have historical evidence that can paint a picture of what is likely to happen down the road when you get to acceptance testing. Are you planning on time in your schedule to fix all these defects during acceptance test?*

**Requirements Engineer**: *No, because we were told to assume acceptance testing would go smoothly, but I've never seen that happen and I've been doing this for twenty years.*

**Analyst**: *Do you know what Ben Franklin said the definition of insanity is?*

In this scenario note that the requirements engineer has both contextual data and objective data, but he appears frustrated and has trouble communicating the facts and getting the timely action needed.

# The Problem and How Lean Six Sigma Can Help

The problem described in the preceding story is that the Requirements Engineer observed a large number of requirement defects, but the analysis didn't go deep enough and the data didn't get specific enough *before the organization started taking their measurements* to help them make the right decision. So how do you ensure you have done sufficient analysis when you are setting up your measurements? This is an area where many organizations could find practical help in the Lean Six Sigma tool-kit.

The Lean Six Sigma methodology includes many proven techniques including techniques to help you set up a measurement system with the right measures given the problem you are trying to solve.[27] When setting up a measurement system using the Lean Six Sigma DMAIC[28] [42] method the importance of selecting measures that clearly tie to your points of pain that can provide high value to your customer, if solved, is emphasized.

Not taking the time up front to ensure you have done enough analysis to narrow down those pain points is where many organizations go wrong. As an example, sending someone to general requirements training wasn't the appropriate response in this scenario. Digging deeper and stratifying[29] [43] the data (data problems, ambiguous requirements) got us closer to the real root cause. But even as we are getting closer to the root of the problem, this new and more specific information will not improve our performance unless appropriate and timely actions are taken.

---

[27]Techniques to help when setting up measures include cause-effect diagrams, brainstorming and use of the 5 Whys technique. The 5 Whys technique is discussed later in this book in Chapter Twelve.

[28]The Lean Six Sigma DMAIC method is a five phase (Define, Measure, Analyze, Improve, Control) method to improve the performance of an existing system. For more information refer to https://en.wikipedia.org/wiki/Six_Sigma.

[29]Stratifying your data is a recommended Lean Six Sigma technique when setting up measures. Stated simply, stratifying your data means to break it into groups more relevant to the problem you are trying to solve. For more information refer to http://www.goleansixsigma.com/stratification/.

As we saw, our requirements engineer was getting quite frustrated answering the analyst's questions and didn't know how to proceed even though he had good information. So, what else could he have done? This leads us to a technique I refer to as a *Structured Real Story* [31, 32].

# Putting the Pieces Together With a Structured Real Story

The more specific we can be with our data the better the chances that we can understand and communicate the right resolutions that can help performance. But we have to be able to communicate the importance of that resolution to the right people at the right time with reliable accurate data to gain their support for timely actions.

One way to do this is through the use of what I refer to as a *Structured Real Story* [30] with graphics displaying both the objective and contextual data together. This technique can be used informally by practitioners, or it can be used together with the CMMI high maturity practices in a practical way as I will explain shortly.

By *real story* I don't mean a make believe story, but a truthful story that is arrived at through the use of a combination of irrefutable objective data, and supporting contextual information.

An example of the stratified objective data that we now know from the deeper analysis can be seen in Figure 10-3. In this figure the stratified categories of requirements defects are displayed. This information gets us closer to pinpointing the real (and high potential benefit) problem that we heard–data problems and ambiguous requirements.

---

[30]Crosstalk, November, 2006, "Uncommon Techniques to Grow Effective Technical Managers http://www.pemsystems.com/pdf/Grow_Eff_Mgr_Crosstalk_Update2.pdf (published in online version of Crosstalk only)

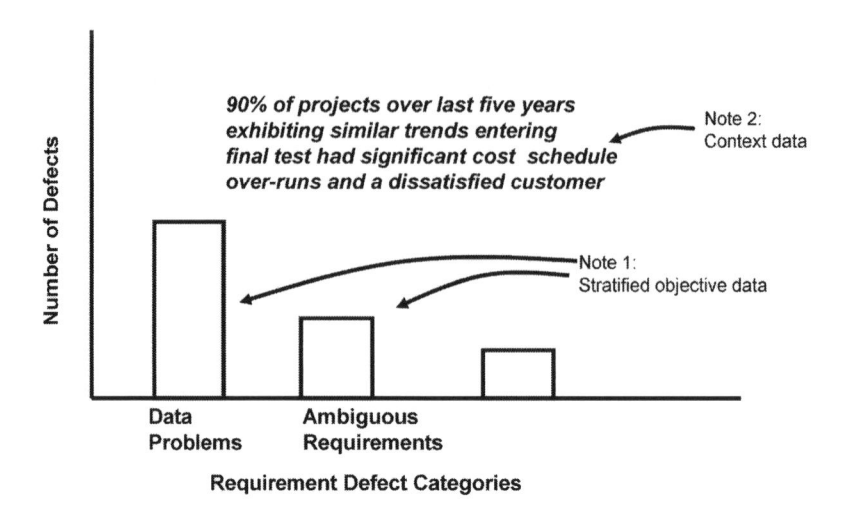

## Chapter Ten—Right Patterns Through Practical High Maturity

**Figure 10-3 Context and Objective Data**

This is more specific and objective data, but it still isn't enough. By itself it doesn't communicate what the right resolution should be, and when that resolution needs to take place to reduce risk to performance. This is where we need the contextual data to tell the story, and that story should be taken as far as possible– including recommendations and consequences if recommendations are not implemented within a specified time window. This contextual data starts with the annotation in Figure 10-3 (note 2), and is continued with multiple annotations of contextual data in Figure 10-4. Together these charts can be employed to effectively communicate what is really going on and what needs to happen to increase the likelihood of this project meeting its performance objectives.

## Sidebar: Objective and contextual data

By objective data I mean data that comes from an outside source providing some level of independence or separation reducing the risk of potential bias. There are two types of contextual data– in the small and in the large. Contextual data in the small refers to data such as "customer A has more data problems than customer B". Contextual data in the large refers to data that results from stepping back and conducting a deeper analysis leading to data such as the annotations on Figures 10-3 and 10-4.

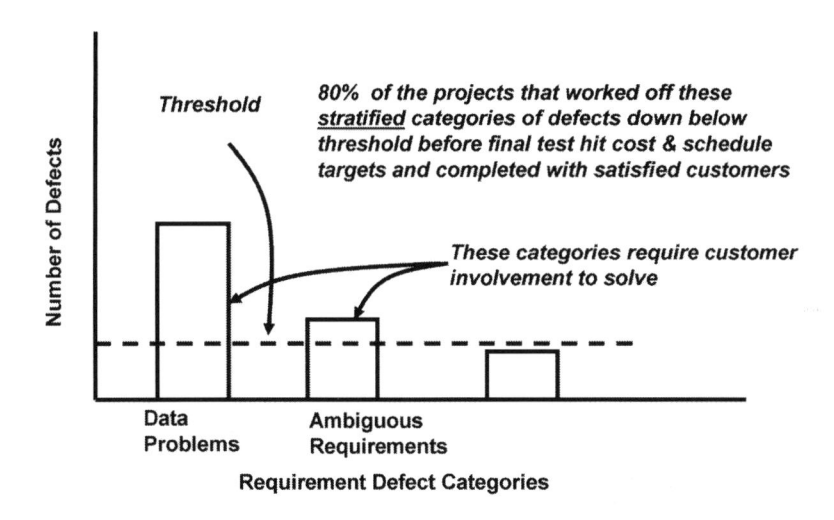

Figure 10-4 Process Performance Model

# Implementing CMMI High Maturity Practices in a Practical Way

Stated in simple language, the purpose of process performance baselines and models– which are essential to the CMMI high maturity practices– is to help individuals and organizations use data effectively to improve performance.

The CMMI defines Process Performance Baseline (PPB) as a "documented characterization of process performance which can include central tendency and variation." Simply put, process performance baselines provide data indicating how your team performed in the past. I view both the stratified objective information (e.g. data problems and ambiguous requirements) and the contextual data (e.g. annotations) as part[31] of the process performance baseline.

The CMMI defines Process Performance Model (PPM) as a description of the relationships among the measurable attributes of one or more processes, or work products developed from historical process performance data and is used to predict future performance. [44]

These charts together provide an organizational performance model in a critical area that has hurt this organization's performance in the past. They demonstrate "relationships among measurable attributes" by explaining the pain these kinds of defects have caused in the past with respect to achieving organizational objectives. They also demonstrate likely consequences in future performance if certain actions are not taken.

Following is the voice track that the Requirements Engineer could use to communicate the story effectively to a senior manager, or stakeholder, who he or she needs to get buy-in from for timely action that is required to sustain project performance.

## Requirements Engineer's Story Voice Track

*"As you can see from my first chart we are experiencing certain types of requirements problems, and as you can see from this chart, 90% of all our projects over the last five years that have experienced these same trends and went into final test with this level of defects ended up with significant cost and schedule over-runs and dissatisfied customers. On the other hand, as you can see from my second figure, of those projects that have been able to work those defects numbers down below the identified threshold, 80% ended up with very satisfied customers, and achieved their cost and schedule targets.[32] I need to get my customer counterpart in here to work these issues*

---

[31]From a CMMI perspective a PPB must go beyond just bar charts to include both central tendency and variation (spread). Refer to [2] Glossary page 599-600 definition of "statistical and other quantitative techniques". Statistical process control is discussed further in this book in the following chapter.

[32]Conducting the research to uncover this important contextual data is part of the analysis required in developing sound performance baselines and models.

*within four weeks. We have done everything we can on these defects and without the customer here we cannot solve them. I need your help."*

> ## Sidebar: Traditional process engineer
>
> One of the reviewers of an early version of this book said I was confusing process with performance and that we should keep them separate. This reviewer was a process engineer in a large organization who did not want to get involved with project specific performance issues. This is part of the reason why some organizations are not achieving their performance goals. Too many organizations today continue to artificially separate process (or practice) and performance. As long as this continues you can expect to fall short of your organizational performance goals.

> ## Sidebar: General
>
> To understand how we can begin to solve this problem refer to the next sidebar titled, "Evolving the Traditional process engineer's role" and to the section below titled, "Increasing Practitioner's Involvement with Process Performance Baselines and Models."

The contextual data helps to communicate the right story. The objective data substantiates the contextual data. Objective data alone can be misleading. Contextual data alone cannot be substantiated. I refer to the approach I am describing here as *"Practical High Maturity"* because we aren't just collecting data and creating charts. We are rapidly analyzing the data and using the data to make recommendations that can improve performance today. Practical high maturity means using historical data ( both objective and contextual) to rapidly locate where changes can improve performance and drive those high value changes leading to high value performance improvements.

> ## Framework Vision: Evolving the traditional process engineer's role
>
> A primary goal of our framework is to support professionals using and adapting practices and patterns, not just defining them. This means integrating practice/process improvement with practice usage. Nevertheless, not all practitioners have the competency, or desire, to develop or even modify practices and patterns. Furthermore, for practitioners to use the CMMI higher maturity practices as I have suggested in this chapter will often require significant analysis to gather the appropriate contextual and objective data. These are areas where the traditional process engineer's role can evolve helping practitioners perform more effectively each day on the job.

# Key Characteristics of Structured Real Stories

Structured real stories have a number of key characteristics. They always include both objective and contextual data, they should also identify a consequence if action is not taken, and they should provide a recommended plan of action, or actions already initiated.

Note that we have all of these characteristics in the Requirements Engineer's Story. Refer to table below.[33]

| Key Characteristic | Example |
|---|---|
| Objective data | Number of defects per category |
| Contextual data | 80% of projects that worked down.... |
| Consequence if no action taken | Schedule overrun, dissatisfied customer |
| Recommended action | Get customer in within 4 weeks |

---

[33] If you decide to use the Structured Real Story Technique it is recommended that you add a statement of the problem. This will help you recall the context when referencing the story later.

# Increasing Involvement with Process Performance Baselines and Models

Contextual data collection and use is a weak area in many organizations, including high maturity organizations, and is often found to be a major contributor to falling short of performance goals. Part of the contextual data for your Structured Real Story can be built from notes acquired through informal communication. Other parts, such as data on past trends, may require deeper research and analysis of historical data. The objective data should come from sources that gives us high confidence in its accuracy, such as organizational cost and staffing data– but it does need to be stratified (broken into more relevant groups to the problem you are trying to solve) to support your specific context as discussed previously. We want the source of both types of data to be as reliable as possible minimizing the likelihood of bias.

The Structured Real Story technique can be viewed as a technique to help practitioners use performance baselines and models in a practical (and possibly less formal) way by helping them integrate the high value improvements needed in a timely way. Refer to Figure 10-5.

## Sidebar: Motivation for increasing the involvement of practitioners in the development and maintenance of process performance baselines and models

Some may question why you should increase the involvement of practitioners in the development and maintenance of process performance baselines and models. The CMMI for Development Guidelines provides a hint where it states that objectives "can be established at multiple levels of an organization" and that "one approach that aligns the whole organization toward achieving the organization's...objectives...is Hoshin Planning, or Hoshin Kanri. Hoshin Planning is a strategic planning methodology popularized in Japan in the late 1950's by Professor Kaoru Ishikawa where each person is the expert in his or her own job. Authority is delegated as much as possible, humanity is respected, and individuals are held accountable. Another hint provided by the CMMI guidelines tells us, "Even when organizational PPBs are established, individual projects may still benefit from establishing their own individual PPBs when they have accumulated sufficient data. In today's ever-changing world this hint can help motivate the need for the rapid feedback and improvement cycle we need to achieve and sustain high performance.[45]

**1. Problem Statement**

**4. Recommendation**

**2. Objective & Contextual Data Gathering**

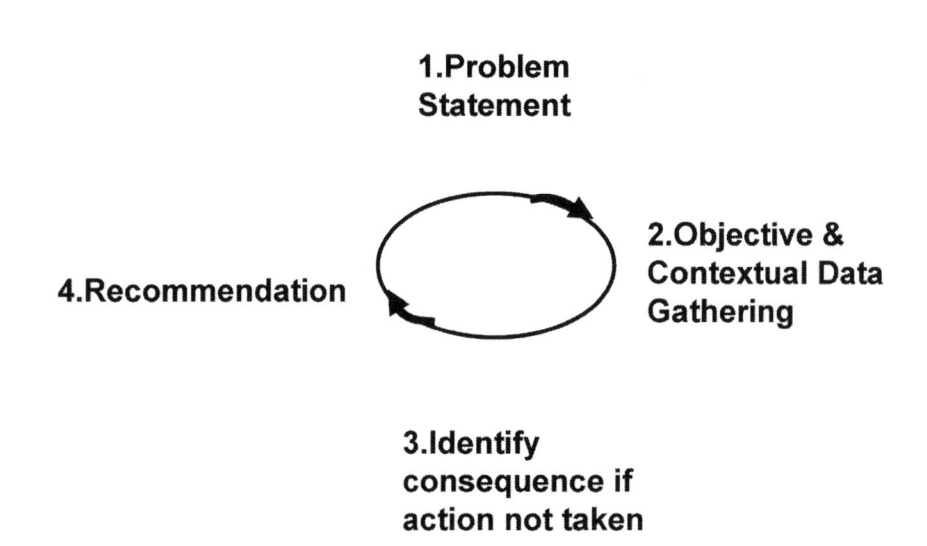

**3. Identify consequence if action not taken**

Figure 10-5 Structured Real Story Technique

You need both the objective irrefutable facts, and the less formal information which provides a necessary context for the "story". Let me now give you a real business example that can help to further motivate the value of the Structured Real Story technique.

# Business Example Motivating the Structured Real Story Technique

One of my clients who utilizes my services to train their project leaders asked me to sit in on a few of their senior management project reviews. They wanted me to update the project leader training, but felt I needed to observe first hand what was currently happening in the Senior Management Reviews.

The first observation I made was that the senior manager who was suppose to be receiving the briefing was driving the meeting. This happened because the project personnel provided the senior manager with the briefing material before the meeting, and then the senior manager came to the meeting with a list of questions. As a result there was no real "briefing" being given. The project personnel just sat back and tried to respond to each of the senior manager's questions.

As it turned out the data they provided in the package was quite good from an *objective data* perspective. It included the required schedule, staffing, list of risks, and cost expenditure information. But despite this the team came across as being extremely unprepared for the meeting.

The reason, in my view, was because they hadn't prepared for the meeting by also collecting the *contextual* data needed to provide all of the rationale and related factors that had led to decisions that had been made. As a result their answers to the senior manager's questions were often weak, and left them with actions to follow up on when they should have had all that information at their fingertips in the meeting.

> ## Sidebar: Structured Real Story helping with competencies of agile team members and teams
>
> Although the intent of Structured Real Stories is to help you with decisions when the stakes get high, this technique can also be used by team members in organizations using an agile approach to help them prepare for daily standup meetings. With self-directed teams it is the responsibility of all team members to manage their own tasks, and speak up when they hear something from teammates that is inconsistent with the project or current sprint agreed goals. Too often we see agile daily standups where team members just focus on their own tasks and tune out when their teammates are speaking. The Structured Real Story technique is one way to help team members rehearse their practices to be better prepared for daily standups. This is an example that demonstrates how helping individuals to become more competent team members helps the overall team competency. Team competency– which is what we ultimately all want–is achieved by helping team members one at a time.

What they had failed to do was to prepare ahead of time by playing devils-advocate, and thinking-through the questions they should have known were coming. This is what the process of building a Structured Real Story is about. It helps you practice your practices ahead of time so you are ready to respond to either a senior manager's question or to a real situation on your project where you don't have a lot of time, but you need to make a rapid decision. Where this really hits performance is when the preparation leads to a recognition that a different decision might actually be in order based on considering both types of data in a logical fashion.

If my client had developed this level of information before his meeting it would have made most of the senior manager's questions straightforward to answer. Building a Structured Real Story can help you get the right data fresh in your mind when you need it most to help you make the best decisions under your project stress, or answer on the spot the tough questions a Senior Manager, coach, or teammate is likely to ask. More importantly, if they had needed management help, not having properly prepared would have meant a great opportunity lost!

In the next chapter we continue our investigation of high maturity practices, digging deeper into the subject of statistical process control, and how these practices can help practitioners and teams that want to increase their agility and achieve and sustain higher performance.

## Chapter Ten Summary Key Points

- Often when a small problem is misdiagnosed due to inaccurate data it leads to an amplified chain of events that only adds to poor organizational performance.
- Once you have achieved your performance objectives you have to keep making small and specific changes to address the changing world around us, if you want to sustain that performance.
- Context data collection and use is a weak area in many organizations, including high maturity organizations, and is often found to be a major contributor to falling short of performance goals.
- Without contextual data, objective facts can be misleading. But contextual data without objective facts cannot be substantiated.
- The more specific we can be with our data the better the chances that we can understand and communicate the right resolutions that can help performance.
- Charts annotated with appropriate contextual data can provide a powerful organizational performance model demonstrating critical weaknesses and specific resolutions needed.
- Structured real stories always include both objective and contextual data, they should also identify a consequence if action is not taken, and they should provide a recommended plan of action, or actions already initiated.

# Chapter Eleven– Practical High Maturity and Agile Retrospectives

*"We don't want just any theory. We want a theory based on mathematics."* Bertrand Meyer, SEMAT Kickoff Meeting, Zurich, March 2010

There has been considerable controversy over the value of using rigorous mathematics, including statistical process control, when it comes to software engineering. An example of this controversy occurred on day one of the SEMAT[34] kickoff meeting in Zurich, Switzerland in March, 2010 when Bertrand Meyer stated:

*"We don't want just any theory. We want a theory based on mathematics."*

Alistair Cockburn responded by raising concerns about a mathematical basis given the lack of precision in our current understanding of software engineering.

So just what kind of a theory makes sense for software engineering? The SEMAT vision includes the following statement:

*The kernel shall rest on a solid rigorous theoretical basis."*[35]

And the vision calls for:

*"The identification of specific theories or theoretical areas that hold potential for SEMAT, backed by examples of their successful application to specific software engineering practices."*

During the discussion on theory and software engineering at the Zurich meeting Ivar Jacobson quoted Kurt Lewin when he said,

*"there is nothing more practical than a good theory."* [36] [46]

---

[34]SEMAT stands for Software Engineering Method and Theory. For more information refer to www.semat.org or Part III of this book.

[35]"The term "kernel" refers to a set of essential elements common to all software engineering efforts.

[36]http://www.selfdeterminationtheory.org/SDT/documents/2006_VansteenkisteSheldon_BJCP.pdf

Could statistical process control be one of those practical theoretical areas that hold potential for SEMAT and software engineering based on its successful application to software engineering? [47, 48].

In this chapter I explain in a simple way how statistical process control, which provides the fundamental theory underlying both the CMMI high maturity practices and the Lean Six Sigma DMAIC methodology, can be applied by practitioners in a practical way today together with Agile retrospectives to help organizations achieve and sustain higher performance. I also explain where many organizations that have used statistical process control in the past have gone wrong costing them the high value potential performance improvements sought.

## Statistical Process Control (SPC) Simplified

To understand the theory underlying statistical process control one needs to first understand the theory behind control charts. Due to the associated mathematics, some practitioners are quickly turned off when first introduced to the idea of control charts. But today we have tools to handle the related math, and therefore we don't need to let that become a stumbling block to understanding. Furthermore, we can discuss and understand the theory underlying statistical process control– which is actually quite simple and practical– without delving too deeply into the mathematical side.

Another reason why statistical process control has been controversial when it comes to software engineering is because of how organizations have applied it in the past failing to gain the promised payback. In the vast majority of situations I have observed, this failure has little to do with the theory of statistical process control. Rather it is primarily caused by a failure to set up a measurement system where the right measures are collected, analyzed and acted on in a practical and useful way.

By the "right measures" I mean measures that lead to what is critical to quality in the eyes of the customer and/or the development team. By "analyzed and acted on in a practical and useful way" I mean looking at the data and acting on it, from the perspective of looking for actionable solutions that could provide high potential value to your customer and/or team in the reasonably near future.

Control charts, when used as intended, can provide a useful graphical representation of how people are performing a process in comparison to past performance. However, if the historical data that is monitored fails to reflect the real process being used by your people that is causing the most critical pain points in your organization and with your customer, then the effort you expend monitoring that process is wasted and will lead to no (or minimal) performance improvement. This is what I have often observed when high value performance payback is not achieved.

To understand how you can use control charts in a practical and valuable way I have found it can be useful to view your process from two perspectives. The first perspective is a high level perspective that focuses on the results of your process. The second perspective focuses on the activities performed by your team in carrying out the process.

> ## Framework Vision: Representing practices using the framework
>
> When using our framework practitioners will be able to view their practices from two perspectives– results or activity. The results perspective focuses on the goal. The activity perspective focuses on what the team is doing to achieve their goal. The rationale for two perspectives is based on the observation that different organizations like to view their practices in different ways and our vision is to support what works for each organization–not to force them to represent their practices in a way that feels foreign to them. A second rationale is the observation that when teams discuss their practices from the perspective of what the team is doing often insights arise that can help the team improve. These insights might never surface by just focusing on results.

Statistical process control theory says you first work to establish a stable process which means that special (also referred to as "assignable") causes of variation have been removed. An example of a "special cause" is a performer with inadequate training. Then you monitor your process against control limits that are based on what is referred to as the "voice of the process." This simply means the control limits are determined by your own development team's historical performance.

Statistical process control helps you by giving you rules (referred to as run rules) to help you identify non-random variation. Non-random variation is where you want to focus your attention first because this is where you can help your performance without changing the process. Another way of saying this is through root cause analysis we investigate outlier situations to determine if there are assignable causes.

Assignable causes are very useful to investigate because they represent areas that are likely to be hurting your performance and we can take action to address them without changing the process. An example of an assignable cause could be providing additional guidance to an inexperienced worker as we saw in the procurement case earlier in the book.

By using statistical process control during a project we can catch potential problems early and take action before they do much damage. If the variation observed is having a positive effect the team could in this case decide to change the process to encourage the continuation of the behavior.

Now, getting back to my point about the usefulness of viewing your process from two perspectives – it is the higher level results oriented process that I have found is best to monitor using control charts. These charts and the run rules can help us identify areas to investigate.

Many organizations understand the statistical process control theory, but what is too often missed is that the results oriented measures I am referring to should also be measures closely tied to the pain experienced by your customer and/or your team that will payback high dividends when solved. In other words, we should be looking at results where we know there is a need for improvement that is worth investing in to improve from a business perspective– and we know we can do something about it in the near future.

Statistical process control is a practical and proven approach to help organizations not only sustain, but also improve their performance. Let's now look at a couple of examples to better understand where organizations go wrong with statistical process control, and how you can avoid similar mistakes. These scenarios will also help you understand why it is valuable to view your processes from two perspectives, and they will help you understand who in your organization should have the authority to make changes to the processes related to the second perspective – the activities performed by your team.

# 1st Example of SPC: The Frustrated Process Engineer

My first example is a common scenario I have observed in multiple organizations. Company A decides they are going to use statistical process control to monitor and control certain processes in their organization because they have a dissatisfied customer. It is clear to them why the customer is dissatisfied because they continually miss their commitments to deliver products on schedule. They have heard that using statistical process control can help them isolate the root cause and put changes in place to improve their performance.

The first decision they make is to have their lead process engineer and their SEPG (Software Engineering Process Group) monitor *defects per KSLOC* (thousand source lines of code). They choose this measure because it is easy to collect and it is a measure that other organizations have reported using successfully to achieve and sustain higher performance. [49] They set the control limits on their control charts using their historical data and the rules that are called for when using control charts (3 standard deviations).

As they gather the data each time a data point falls outside their control limits they perform a root cause analysis. In the first case they find that the reason for the excessive defects is an inexperienced programmer that is unfamiliar with the area of code he has been assigned to work on. The lead process engineer recommends to the project manager that a more experienced programmer should be assigned to assist, but the programmers in Company A are already in short supply and so no action is taken.

In the second case investigated that falls outside the control limits, the lead process engineer discovers through her root cause analysis that the reason for the excessive defects is a poorly designed section of code, and so she recommends that instead of continuing to patch this bad code that it be redesigned or refactored to improve its maintainability. But the project manager tells the process engineer there isn't any spare budget available to conduct any redesign or refactoring and so no action is taken.

In the third case investigated that falls outside the control limits, the process engineer learns that the reason for the excessive defects is poorly written requirements that caused the programmer to misinterpret what the customer needs. This leads to unplanned rework. The process engineer also discovered during her root cause analysis of the requirements problem that the requirements had not been properly peer reviewed and so she recommends remedial peer review training. This additional training was approved by the project manager and conducted, but unfortunately there were no measurable improvements in scheduled product deliveries to the customer that could be attributed to this training.

I have talked to a number of process engineers in multiple companies who have experienced a similar scenario, and similar frustrations where they follow what they believe to be the prescribed statistical process control theory, but find it leads to minimal- or no- performance improvement as was the case in this common scenario. Refer to Figure 11-1.

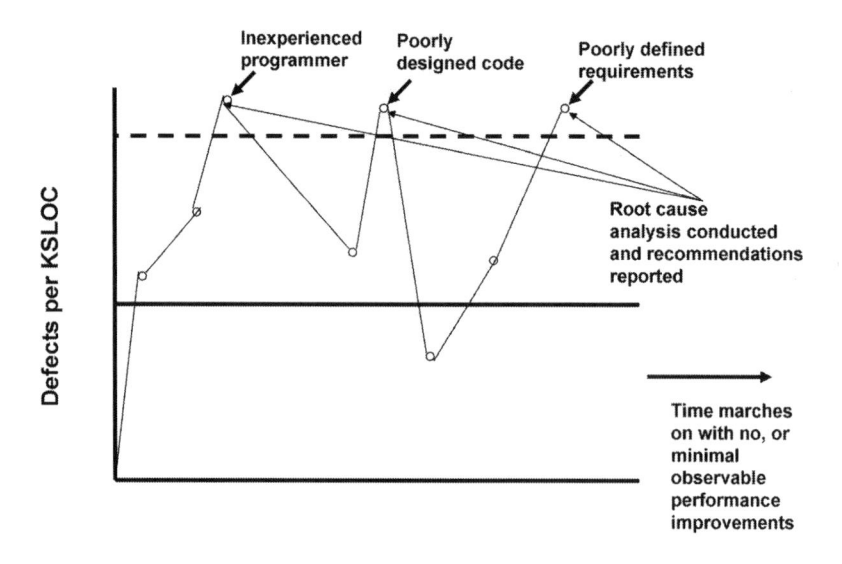

Figure 11-1 Company A Statistical Process Control

# 2nd Example of SPC: Empowering Your Scrum Team

Now let me give you another scenario. Company B decides it is going to use statistical process control because– similar to Company A– they are having trouble meeting their customer schedule commitments. But Company B is also using Scrum [50] and this organization decides to empower its Scrum teams in the implementation of its statistical process control initiative.

The first team to apply statistical process control in Company B is using four week sprints[37] with retrospectives[38] at the end of each sprint. At the end of the first sprint the team discovers through their retrospective that what is hurting the team's performance is the fact that the product backlog[39] items are not being properly prepared by the product owner[40].

As a result the team decides to monitor their *story point efficiency*. Story point efficiency they define as the ratio of how long their team estimates it should take to complete a story to how long it actually takes to complete the story.

The team knows from their own experience that when backlog entries are properly prepared they have been able to achieve a story point efficiency on average of fifty percent. But on this project they know they are not performing this well. Because the team knows they are having trouble with their story point efficiency they decide this would be a good measure to place under statistical process control.

The threshold based on the statistical process rules and their historical data is set at 20%. This will trigger the team to conduct root cause analysis when their story point efficiency drops below this threshold.

In the first case they investigate that falls below the threshold the team finds there is missing data that is needed for test.

---

[37] The heart of Scrum is a Sprint, a time-box of one month or less during which a "Done", useable, and potentially releasable product Increment is created .

[38] The Sprint Retrospective is an opportunity for the Scrum Team to inspect itself and create a plan for improvements to be enacted during the next Sprint.

[39] The Product Backlog is an ordered list of everything that might be needed in the product and is the single source of requirements for any changes to be made to the product.

[40] The Product Owner is responsible for maximizing the value of the product and the work of the Development Team.

They alert the product owner of the need for this test data, and they also marked up a checklist to help the product owner recall the next time to make sure any required test data is included when he prepares the backlog.

In the second case the team discovers a backlog item that wasn't properly broken down and estimated to fit within a single sprint. In this case they alert the product owner and remind him that he needs to work closely with team members when preparing the product backlog so that the team understands the extent of the work and can complete it within a single sprint. The team also adds another reminder to the checklist for the product owner to help him catch similar situations in the future.

In the third case that falls outside the control limits the team discovers that a related risk had not been logged and properly assessed. This hurt the team's story point efficiency because it required the assigned developer to conduct unplanned analysis and report and communicate the risk. Refer to figure 11-2.

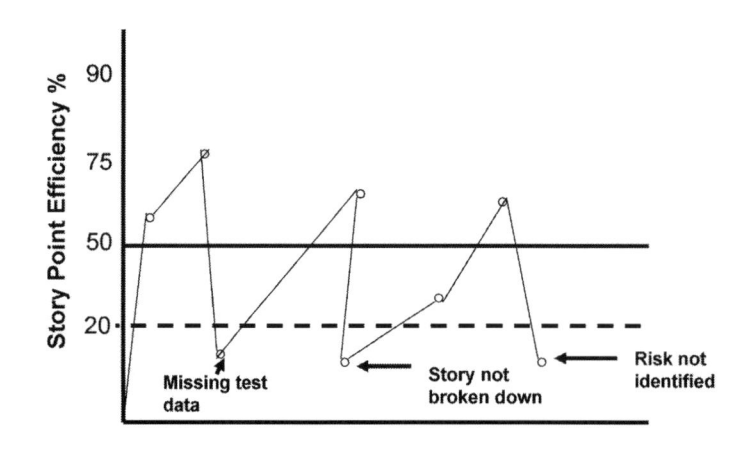

Figure 11-2 Company B Statistical Process Control

In each case investigated an unprepared backlog item led to unplanned delays and waste. The team used this data in their next retrospective to explain to other teammates what they had found and to explain what actions they had already taken and further actions they were recommending for the next sprint. The full team overwhelmingly agreed that the product owner checklists needed to be permanently updated with the suggested improvements to help catch similar problems sooner in the future. These improvements were immediately implemented for the next sprint.

Because the team started with a known problem to solve it led them to choose a measure that connected directly to the past poor schedule performance observed by the dissatisfied customer. As a result, the improvements put in place by the team led directly to noticeable improvements in schedule performance by both the team and the customer.

## Comparing Company A and B's Approach and Results

In the Company B case the process that was being monitored was chosen by the team because they knew it was an area that was causing themselves and the customer pain. The story point efficiency measure that was chosen is what is referred to as a *leading edge measure*. This means it is a measure that leads the team to ask why and leads them to institute changes that will help prevent similar problems in the future.

In the Company B case updating checklists was part of the process because the process included retrospectives and improving their own process. In other words, process improvement was included as part of the responsibilities of the team.

In the Company B case the team was not frustrated by having to go to a project manager and convince him of the need to change. Changing the process was part of the process and therefore the team did not need to explain their actions and get formal approval outside of their own team.

In the case of Company A there was little high value performance improvement observable as return on investment. In my experience I have often found that the Company A scenario plays out at least partially because the process engineer isn't viewed as part of the team and therefore her recommendations are often not given high priority.

Another reason that the recommendations coming from the process engineer may not have been accepted was because the measures had not been *stratified* and therefore the related recommendations were not specific enough to convince the project manager that high value performance improvements that the customer would notice would result.

If the data you monitor is not properly stratified you can spend a large amount of effort chasing outliers that are not the outliers causing your pain. *Following a thread*, as discussed earlier in the book, is one technique that can help you stratify your data, as our Scrum team did in Company B, eliminating wasted effort and helping your team focus their improvements where the greatest payback lies for your investment.

## How an Unstable Process Could Help Your Team's Performance

When using statistical process control a stable process is defined to be a process where all assignable, or special causes of variation have been removed. But is a stable process really best for your performance?

Common cause variation is predictable variation. It is part of the normal variation of a stable process. Assignable, or special cause variation is unpredictable. It is a signal that variation exists that needs to be investigated. The majority of projects that get into trouble do so NOT because of predictable variation, but because of unexpected situations that need to be investigated as soon as they occur so that action can be rapidly taken before significant damage occurs.

When you look at it this way an unstable process (e.g. one that is triggering assignable causes that need to be investigated) could actually help your team's performance because it is alerting you where you need to investigate and make corrections to keep your endeavor on track.

But to really gain the performance benefits you need to have processes that support rapid investigation of assignable causes and taking action to resolve them in a timely way. This is exactly what agile self-directed teams do through their retrospectives and this is why it makes sense to put your practitioners in charge of handling the assignable causes that result from your statistical process control efforts.

When you include improving the process as part of your process you empower your team to continually improve without the need for excessive bureaucratic oversight. Furthermore, you get improvements where they can help the most keeping your team's performance at the high level you need.

# Chapter Eleven Summary Key Points

- The reason that many organizations have failed to gain the promised payback from their statistical process control investments comes down to a failure to set up their measurement system with the right measures.
- Leading edge measures lead a team to ask why and leads them to institute changes that will help prevent similar problems in the future.
- When process improvement is considered part of the process improvements happen in a natural way. They are not viewed as something extra.
- If the data you monitor is not properly stratified you can spend a large amount of effort chasing outliers that are not the outliers causing your pain.
- When you include improving the process as part of your process you empower your team to continually improve without the need for excessive bureaucratic oversight.

# Chapter Twelve– The Real Root Cause and Solution

*"Failure is simply the opportunity to begin again this time more intelligently."* Henry Ford

From an analysis of my personal improvement project notes it appeared I was convinced I was learning the same lessons over and over. I now know I wasn't learning any lesson. I was just gathering more data that told me what I already knew. That is, until I started to collect different and more objective measures that led to a deeper analysis and understanding of my positive and negative patterns. In 2011 I discovered the reason my fixes weren't fixing my problem was because I hadn't yet discovered the nature of repeating specific weaknesses and the impact they had on how one had to tackle causal analysis and resolution.

When I went back to read my notes in 2011 what I found led me to ask: Why were the fixes not permanently sticking? Then I asked: What is different when my swing starts to fail? Am I losing my focus? If so, why? Was something outside my control distracting me? If so, why? Why wasn't I able to solve this problem once and for all? Why? Why? Why?

From a business perspective this is comparable in the procurement case to where we started collecting different types of data and digging deeper asking questions to understand the different situations where the procurement specialist was having difficulty.

On the personal improvement project my next step was to refine my data to better understand the conditions when my swing failed. Back on the business side, in the Requirements Case Study this is comparable to our breaking down the Requirements defect category into the more specific categories of Ambiguous Requirements and Data problems.

# The Real Root Cause on the Personal Improvement Project

The following is taken from my personal improvement project notes dated, Saturday, July 9, 2011.

*Hit large bucket of balls at driving range. Hit ball great with all clubs! Hit twenty straight drives all perfect! Went straight to golf course. Wind blowing left to right on first hole. Hit lousy drive—worse than any of the twenty hit at the range. Next hole hit good drive. Third hole hit lousy drive again. Hit two more drives off third tee—both poor! What the heck is going on! Why can't I swing the same way on two different holes? This makes no sense to me. Or does it?*

It was at this point I started to figure out what was happening on the personal improvement project. While my second level checks had helped, they didn't help in all conditions.

*The real root cause was that I hadn't been practicing under the varying conditions that different holes and different weather conditions presented. In other words, the process I was using was effective under basic conditions, but it broke down when I faced harsher conditions which are a normal part of the game of golf.*

It was easy to hit the ball well at the driving range because the conditions were always the same. On a golf course we have to deal with different wind conditions, narrow fairways, side hill lies, water hazards and so on. There are 18 different holes on a golf course each presenting its own unique challenge, and stresses on your performance.

In business on the procurement case this is comparable to where we started to realize how overwhelming the problem might be with the 100's of different possible types of missing information on the requisition form.

As I analyzed this information I realized this seemed to make sense, except for one thing. If this were true, why had I been able to sustain my performance for a full week with my friend in 2008 where we played different golf courses each day which I had never played before?

It was then I recalled that in 2008 I knew I would be playing golf courses I was unfamiliar with and this would stress my performance. I also knew that if I tried to remember too many different things it would only add stress and probably do more harm than good to my game. I had to put a plan in place that would help me deal with this complex situation, but without overwhelming me. I came up with the following approach.

## The Solution

First, on all golf shots I constantly reminded myself that I didn't need to hit every one perfect. Whenever I sensed the condition[41] on any given hole not feeling right for my game, I immediately backed off and gave myself room for error. For example, in these cases I would shoot for a spot just short of the green, as long as there was no visible trouble there.

Second, whenever I faced a hole with a cross-wind or side hill lie I decided I would take an extra club, choke down on the club shaft for added control and focus on keeping my balance.

Third, on all iron shots into greens I would not shoot directly at the pin, but rather I would pick out a spot on the green that would allow me to miss the shot slightly and still be in good position.

Fourth, during my set up while I was doing the other checks I would also keep aware of my grip pressure. Keeping a light grip pressure was a simple technique I had learned to help me avoid the chain of events leading to my body tightening and fighting itself. Refer to the table below.

| Scenario (or pattern) | Action |
| --- | --- |
| 1 Feeling uncomfortable with conditions | Give room for error(e.g.short of green) |
| 2 Side hill lie | Extra club, choke down, focus on balance |
| 3 All iron shots into green | Don't shoot at pin, Give room to miss |
| 4 All shots | Keep aware of grip pressure |

---

[41]You can see another example of similar thinking in the work Scott Ambler has done in describing the Software Development Context Framework (SDCF). Refer to http://disciplinedagiledelivery.wordpress.com/2013/03/15/sdcf/. This framework calls out scaling factors such as team size, geographic distribution, regulatory compliance, domain complexity, technical complexity, and organization distribution. The idea is that the SDCF can help you to understand the situation you find yourself in, and then tailor your strategy to address the DAD process goals accordingly.

Keeping these four patterns fresh in my mind took pressure off my game and allowed me to make good swings even on holes where I was uncomfortable[42]. They also were reasonable given my goal. I knew I could afford to make a few bogeys[43] and still achieve my goal of having a reasonable chance of breaking 80 each day. I had a strategy to deal with the conditions I would face and I went over that strategy in my mind each day before we played, and often between holes as well so I would be ready to apply each strategy at the proper time. Because there were only four patterns I was able to keep them all in the forefront of my mind quickly recalling what to do when faced with each situation. This strategy felt right from the big picture perspective. It was a plan I knew I could execute.

Abstracting four patterns that I could then review and keep in my mind to improve my performance is comparable on the business side to the five patterns we developed for the procurement specialist. Note we are keeping the number of items to focus on between 3 and 7.

As I recalled this plan from 2008, I realized if I was to sustain my performance over much longer periods of time I would need to create a similar strategy before I went out to play each round of golf that fit with the conditions I would face and my current objectives.

From the business side, you may be thinking at this point that this seems like a lot of work to plan, or create a strategy each day at work. However, this activity does not need to take a lot of time. The reason many people fail to do adequate planning is because they hold the false belief that planning must be a labor intensive task.

Getting back to the personal improvement project, I knew I needed to make sure I was setting targets that continued to help me stretch toward goals that would raise my performance, but not goals that would be so stressful as to lead to poor decisions. In other words, I now realized I needed to be developing more specific plans on short cycles. I now had a strategy to help me perform well under the varying conditions I would face each day.

---

[42] Note that the number of patterns focusing on is between three and seven
[43] A bogey in golf is one over par.

> ## Sidebar: CMMI tip on objectives
> A sidebar tip in the CMMI guidelines states that objectives should motivate superior performance...setting the bar too high may demoralize rather than motivate.

# Insight Into Repeating Specific Weaknesses

In both the business procurement case and personal improvement project we never pinpointed a single root cause that could permanently fix the problem.[44] This is the nature of repeating specific weaknesses. There is no simple quick fix. The fact that there isn't a single resolution is an important insight into repeating specific weaknesses. What this means is that the solution requires a never-ending process of contextual and objective data gathering, followed by analysis and refinement of resolution to maintain performance within acceptable limits.

As an example, in the procurement case the five categories we created to help the procurement specialists were not perfect, and they will need to be refined as the work environment changes over time. In my personal case the four scenarios helped during my marathon week of golf, but by themselves they will not ensure that my performance can be sustained. I must keep taking notes, analyzing the notes and making small changes which I find lead to small improvements each day.

I have concluded from my personal performance improvement experiences and from working with multiple business clients looking for higher sustainable performance the following fundamental.

---

[44] For those who would prefer to locate a single root cause, perhaps it could be viewed as the lack of continuous problem ownership. The resolution is for individuals or teams to own the problem and refuse to stop working it even when it appears they have met their goal.

## Fundamental Fourteen:

The most valuable performance improvements often involve situations that seem to defy resolution because there is no quick fix, but we know they are critical to our performance and we know there is no way to work around them so we must continually deal with them head on.

The situations that seem to defy resolution must be continually dealt with head on because they tend to occur at the most inopportune time hurting our performance just when we need to be on top of our game. Such situations are best handled by continually gathering contextual and objective data, creating the likely scenarios, and practicing at just the right time– like Tom Brady preparing for his next opponent– so you can be ready to make the best decision at exactly the right time in your actual work environment.

Next I will explain a valuable technique to help guide your search for root causes of your most critical performance weak spots.

# Ask Why 5 Times to Guide Your Search for Root Causes

Isolating root causes is not always easy. Plan on continually refining your measures, objectives, and resolutions, and consider asking why at least 5 times. The reason you should ask why 5 times is because often the first few answers just lead to symptoms.[45] [51] Asking why 5 times can help you narrow in on the real root of the problem. But keep in mind that if you are dealing with a stubborn repeating weakness this process will never end, and it is likely you will find yourself studying environmental patterns to aid your performance.

Often when we dig deep asking why, why, why, why, why, performance issues lead to work environment patterns. Besides asking why, ask:

*"What different situations must your people face in their daily work environment?"*

And

---
[45] For more information on the Ask Why 5 Times method, http://en.wikipedia.org/wiki/5_Whys

*"Are there work environment changes you could make that might help your people make the right decision at just the right time?"*

# How Allowing Variation Can Reduce Variation Where it Counts

From a CMMI perspective, high maturity organizations use statistical techniques such as analyzing central tendency and variation (spread).

However, this doesn't mean the goal is always to reduce variation in all areas. Sometimes allowing variation in an individual's personal problem solving approach can help an organization achieve their ultimate goal of fewer defects for the customer, and therefore reduce variation where it counts most.

This is another reason to view your processes from two perspectives as discussed in the previous chapter. It is also another reason why achieving and sustaining higher performance requires that we are constantly attentive to measuring the right things and continually refining our measures on short cycles.

When Tom Brady drove down the field to help the New England Patriots beat the RAMS in the 2002 Super Bowl his first few plays were for short gains. He took what the defense gave him. He didn't try to force it. He was using his rational and emotional brain together in a way that leveraged the accurate patterns he had recently reviewed and stuck in his head before the game.

In my own life, both personal and professional, I learned I must consciously write things down and go back and read them a number of times– even things that sometimes seemed obvious. It took years and many iterations of reading my notes (both on personal improvement efforts and business cases) to recognize the importance of preparing mentally by going over the possible scenarios in my mind so I would be prepared for what I was going to face each day.

It is easy to listen to these seemingly simple lessons and agree. However, it takes commitment, desire and personal discipline to use this information to affect your personal performance. What is easiest is to return to your personal work environment and miss the opportunities for high value improvements you face each day.

## The Right Kind of Practice

To use the techniques described in this book requires practice. They require practice at just the right time with just the right objective and contextual data to trigger the right behavior we need to sustain higher performance. This is what I mean by *"integrated practice"*. Not mindless, brute-force practice, but practice integrated with actual performance conducted at just the right time.

By taking the time to create a *Structured Real Story*, or using the ask why five times technique at just the right time you are preparing in your head for your upcoming real performance. You have to learn to do it in your head, before your body parts can perform in a personal improvement area, or in today's common stressful business work environments.

### Fundamental Fifteen:

Just knowing what is happening isn't enough to keep it from happening again. You must practice continually at just the right time with the right objective and contextual data, if you really want to make the changes necessary to sustain higher performance.

### Framework Vision: An Example

So, is this the end of the story? Not quite. In the next part of the book we present a real thinking framework, and give you some examples demonstrating how this framework can help us solve the problem we all face.

# Chapter Twelve Summary Key Points

- The most valuable performance improvements often involve situations that seem to defy resolution because there is no simple answer, or quick fix, but we know they are critical to our performance and we know there is no way to work around them so we must deal with them head on.
- Often when we dig deep asking why, why, why, why, why, performance issues lead to work environment issues that we must keep constantly aware of.
- Just knowing what is happening isn't enough to keep it from happening again. You must practice continually at just the right time with the right objective and contextual data, if you really want to make the changes necessary to sustain higher performance.

---

# PART III A Thinking Framework

# Chapter Thirteen: An Example–The Essence Thinking Framework

*"The secret of success is constancy to purpose."* Ben Franklin

SEMAT stands for Software Engineering Method and Theory [52, 53]. SEMAT refers to a community of volunteers that developed a new framework that is defined in the Essence Specification [54]. This specification was adopted by the Board of Directors of the Object Management Group (OMG) as an official OMG standard in June, 2014 (www.omg.org). In its simplest form Essence can be thought of as a thinking framework [55, 56]. The SEMAT initiative was started by Ivar Jacobson, Bertrand Meyer and Richard Soley primarily to address repeating problems observed within the software industry, but its potential extends beyond just software.

> ## Sidebar: Note about Essence figures, cards, and definitions
>
> The Essence figures with Alphas, and Competencies and the figures with Essence Cards provided in this part of the book and in the Appendicies to this book are provided courtesy of SEMAT Incorporated and the community of SEMAT volunteers. The cards are an early version and provide abbreviations of the full checklists. For the full and latest version of the checklists refer to the OMG website [54]. For the latest version of the cards refer to the SEMAT web site (www.semat.org). It is also worth noting that some errors were made in changes to definitions (e.g. practice, activity) in the first OMG release of the Essence specification. Requests have been submitted to the OMG to make appropriate corrections.

> **Sidebar: Note about terminology used in this part of the book**
>
> In this part of the book the term "apply" means the same as "use." The phrase "Essence System" means the Essence framework together with any extensions you have defined. The phrase "Essence Approach" means powering whatever your organization is doing today using the Essence framework. The phrase "bring your practices into the Essence system" means representing your practices as extensions to the Essence framework.

I have spoken on the subject of SEMAT at conferences and universities, and each time I have encountered attendees who have questioned the need for a new framework. Therefore in presenting the framework in this part of the book I do so from the context of how this framework can help solve the problem we are all facing in ways that other performance aids have fallen short. As I describe features of the Essence framework I will also relate those features back to our thinking framework needs discussed previously.

But first I want to review the problem we all face discussed in the first two parts of this book, and then I will explain how this new framework can help.

# Review of Problem we all Face

I used the first diagram in the introduction of this book (**Fig Intro-1** CMMI Theory and Observations) at the 2010 Zurich SEMAT kickoff meeting to make an observation based on my years of experience helping organizations in their quest to improve.

I started as a programmer in 1973 and for the first half of my career I was a practitioner– programmer, technical lead, project manager. For the second half of my career I have been on the consulting/coaching side, and during this part of my career I have observed millions of dollars being spent annually on improvement efforts that too often fall short of their goals.

The referenced diagram was actually given to me by a colleague in a CMMI Level 5 organization. The diagram clearly shows there is something wrong with how many organizations are approaching performance improvement.

My position, as I explained earlier in this book is that the theory of performance improvement is not wrong, but we as an industry are not doing a good job of translating theory to practice. Stated differently, there exists a gap today between the theory of performance improvement, and how we implement that theory.

I have also observed common patterns that can help us do better and I will summarize those in a moment, but let me first say a little more about the referenced figure.

If you look at the software professionals in the world we could broadly view them as falling into two groups– practitioners or "doers" and the process professionals who focus on defining or assessing the processes or practices that the practitioners follow. It has been observed that often practitioners are skeptics of written process and process improvement initiatives. One reason is evident in the referenced figure. Stated bluntly, it is the common message from practitioners–

*Don't give me extra stuff to do, if it doesn't help.*

Now let me share some observations that will set the stage for our discussion of the Essence framework.

# Why Are We Falling Short When Millions Are Spent Annually?

I referred earlier to millions of dollars being spent annually on efforts that too often fall short of their goals. It has been my observation that a huge percentage of these dollars are expended on ideas that never reach the practitioners for whom the improvement is intended. One reason for this is the fear that we can't change until we know the change will help.

Therefore, we run "pilot projects", but these pilots are notorious for being setups-to-succeed because they are run in ideal environments that rarely reflect the common conditions practitioners often face (such as shortfalls in staffing, ambiguous requirements, overly aggressive schedules, or stakeholders who are too busy to collaborate). So the pilot project goes well and this leads to the false belief that the new or improved process will help practitioners in their real project situations.

We then deploy the process, with maybe an hour of classroom training explaining how the process works in an ideal world. But the problem with this implementation is that things rarely work on real projects like they do in these controlled "lab-like" environments.

What practitioners really need isn't help with processes that tell them what to do when everything on the project goes right, but rather they need help with how to handle the difficult, but real, situations that often create obstacles to their success.

Now lets talk about another pattern related to how we implement performance improvement.

## What's Really Wrong with How we Implement Performance Improvement?

If you want to improve performance, you first need to identify where that improvement is needed. Then you measure your current performance, then you change something and then you measure again to see if you have achieved improvement. This is sound in theory, but now lets look at the pattern that often happens in implementation.

Instead of first identifying where improvement is needed, we start by gathering large quantities of data, and after a lengthy period of time we finally get around to looking at the data.[46] But because we have so much data now, and because much of it was collected long ago, people often no longer recall the details of what happened. Therefore we can't figure out how to analyze it. As a result we end up stepping back and looking for high level trends of aggregated data; and then we take our best guess at where improvement is needed. This leads to responses such as the following which we saw in a specific scenario described earlier in the book. [47]

> "I got told I had to attend requirements training, but it was general stuff I learned years ago that didn't help with the problems I am facing today."

So where did we go wrong implementing the theory? What I am about to say may fly in the face of what many have been taught.

---

[46]While it is true that the agile movement has helped many organizations change this poor behavior, there continues to exist in 2014 a large percentage of organizations that have not changed this failed approached.

[47]Refer to Chapter Three for a specific example.

The theory many have been taught is that we need to capture a large amount of historical data before we analyze the data. But the problem with this theory is that the speed at which things change today causes historical data to rapidly lose its relevance to what is actually happening. Also because the world is changing fast, by the time we get around to analyzing the data, the people involved often can't recall what happened because they are now on a different project.

Don Wheeler says in his classic book, "Understanding Variation:"

*"As data are aggregated they lose their context and their usefulness."*

And Doug Hubbard tells us in his book[48] [27] that often we can learn a great deal from small samples of data.

So the right place to start isn't collecting large volumes of data, as many of us have historically been taught, but rather to start with small samples.

# Understand the Problem, Carefully Approach the Solution

But now the trick is to figure out where to take those small samples from because if you take them from the wrong place it won't help you really improve.

So how do you figure this out?

If you are thinking hard about this question, you are probably thinking too much because it isn't that tough!

All you have to do in most organizations is just go talk to a few people in the organization. Ask them to just talk about their job and the high potential value pain points usually quickly jump out. And you will learn with little effort about the key areas in the organization where you can really make a difference.

Examples that are commonly heard include:

*We are always working with unclear requirements.*

*The requirements keep changing.*

---

[48]Doug Hubbard's book is titled, "You Can Measure Anything".

*We can't get our stakeholders to work with us.*

*We get pressure to release before we have done adequate testing.*

*The hardware is always late and it cuts into our software test time.*

The next step is to recognize what you now know and what you don't know. You have identified pain points that could payback high value if solved. This is important, especially to get management and practitioner buy-in. But you don't yet know the solution. And you don't want to jump to the solution until you fully understand the problem. This is a mistake too often made which leads to working on symptoms rather than root causes.

The next step is to understand more about the nature of these common pain points that really tend to hurt performance so you can locate the right solution that can lead to high value and lasting improvements.

## Locating the Right Solution

I have found that common pain points– the ones that really hurt your performance the most– have certain characteristics.

First, they tend to repeat, and they tend to repeat at the worst possible times, often right when you need to be performing your best. As an example, when you find yourself at a critical spot in a project and you need to make a key decision. What makes the situation even more difficult is that often in these cases you are under stress and need to make that decision rapidly.

I have done some analysis and have found these types of situations can be partitioned into two types. The first type are those that repeat because they aren't one problem. The second type are those that repeat because conditions mask them.[49] Earlier in this book I discussed the power of recognizing simple patterns (first type) where simple actions could be taken.

In the procurement case (first type) discussed earlier in the book where we demonstrated a small set of simple actions that could be captured in a table format, the following key points were noted:

---

[49]For an example of the first type refer to the Procurement Case study discussed in multiple chapters in Part I and Part II of this book.

- Management and customer observed measurable improvement.
- The situation did not require large quantities of data, but rather deeper analysis of small data samples.
- The right solution included not just collecting and analyzing data, but rapid action to improve, monitor result, and refine action.
- The patterns identified were not perfect. Over time new problems tend to creep in, and therefore practitioners need to take responsibility for maintaining their patterns.

Now lets move on and discuss the tougher second type situations.

# Handling the More Difficult 2nd Type Situations

Not all problems can be broken down as neatly as the first type into simple answers. And when we try to do this with the second type is where we get into trouble. The second type of situations are more difficult because the actions needed are not as straightforward. The answers to this type don't fit neatly into a simple table like the first type.

And when organizations try to take these type 2 situations and make them fit type 1 we end up with large process documents that are not useable by human beings, or we end up with guidance that is so overly simplified that it provides no useful help to practitioners.

Examples include process guidance that says:

*Do requirements before design*

*Peer review your design before you code it*

*Test before you integrate.*

The reason such guidance is not useful is because the real question practitioners often face is not "what should I do", but "how much" should I do.

Often, in hindsight, looking back at troubled projects you can often point to what appears to be small things that weren't handled properly early. I am certainly not the first to make this observation. One of the best known references to this observation was made by Fred Brooks in his classic book, "The Mythical Man-month" [29], where he asks,

*How does a project get to be a year behind schedule?"*

The answer is, "*one day at a time."*

But while we all intuitively know this is true, this observation alone is not very helpful. With this knowledge it doesn't help to tell people to make sure they handle all the small things because not all small things lead to big things, and the real problem is that on the majority of projects there just isn't enough time to do everything. So how does one decide which small things need to be handled now, when there is never enough time to do everything?

To understand the answer to this question and what we can do to address the related problems just described, lets now look at the Essence approach.

## The Essence Approach

Let me start with some general facts about the Essence approach and as I explain them I will connect key points to the problems we've been discussing throughout this book, including how to address both types of pain points. I will also explain how the Essence framework addresses the needs of our thinking framework vision.

In its simplest form you can view the Essence framework as a flexible thinking framework. Consistent with our framework vision discussed earlier– by thinking framework I mean an aid that helps individuals and teams think-through the tough problems and helps them make better decisions related to their specific situation. The Essence approach powers whatever you are doing today by guiding a team's decisions and continual improvements. Specifically, the Essence framework will help a team ask key questions related to the things that are essential for success on all software engineering endeavors.

This is different from other improvement frameworks that only help with specific dimensions of your challenge. For example, Scrum is a management framework that provides assistance helping teams self-manage their work, but it doesn't help with the technical practices related to building and testing software.

Essence is a comprehensive framework that helps your team assess essentials for success and it provides objective information your team can use to help make priority decisions related to what is most important to focus on next. This can help your team gain the needed attention to key issues early that might otherwise be missed.

## The Essence Framework

Some of the key elements inside the framework that I want to focus on in this part of the book include Alphas, Alpha States, Alpha State Checklists, and Competencies.

Alphas are the "essential" things we work with as referenced earlier in the discussion of our thinking framework vision. Alphas have states and checklists. It is the states and checklists that give us the "objective data" that we use to aid our assessment of progress and health. States and checklists help our assessment be as accurate as possible leading us to take appropriate action in a timely way– especially when needed to address those tough pain points.

I will explain in a moment how Essence checklists differ from the checklists used by many quality departments, and by many organizations that use the CMMI model for improvement.

I will also explain shortly how the Essence checklists can help address the complaint discussed earlier we hear from many practitioners:

*Don't give me extra stuff to do, unless it helps.*

But first lets look a little closer inside the Essence Framework.

## The Seven Alphas Inside the Essence Framework

The seven Alphas inside the framework include Opportunity, Stakeholders, Requirements, Software System, Work, Team, and Way of Working. Refer to Figure 13-1.

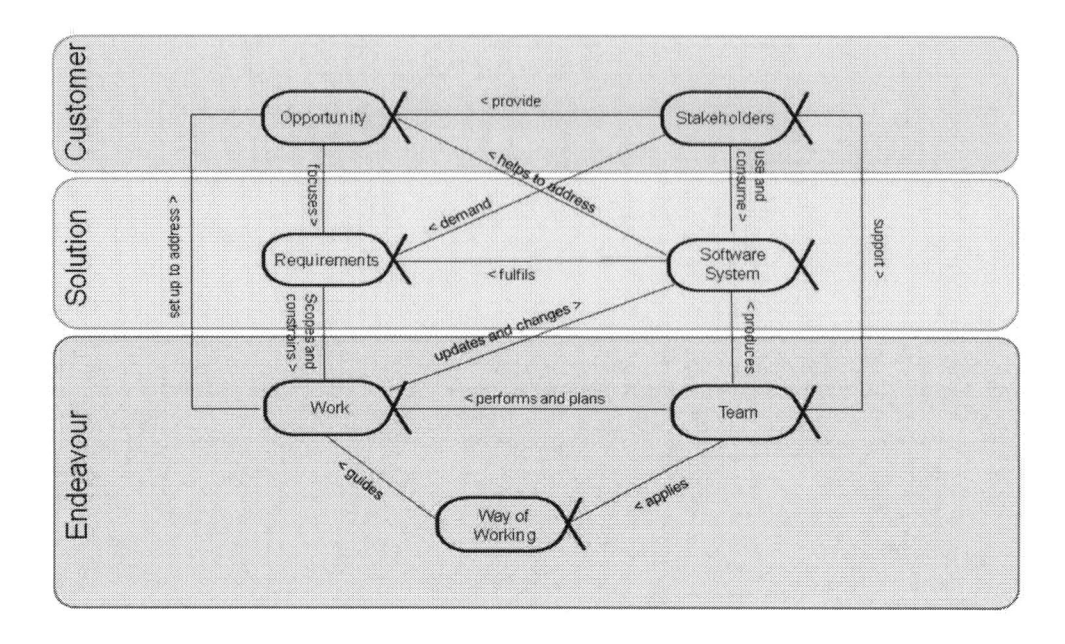

**Figure 13-1 The Seven Alphas Inside the Framework**

The Alphas can be viewed as critical indicators that should be monitored and progressed for endeavor success. For example, Team is an Alpha that needs to be monitored and progressed because the health of your team is always critical to the success of your endeavor. While team members are an important part of the Team, they are also stakeholders. Stakeholders must also be monitored and progressed (both those inside and outside your team) because the health of your stakeholders is also critical to the success of your endeavor. It is the states and the checklists that help you ask the right questions leading to your best possible assessment of where you are and where you need to focus next.

You can use the Essence framework whether you decide to bring your practices into the Essence system, or not. I will explain this further in a moment, but first I want to get back to how Essence checklists differ from checklists used in many organizations and how they can help your team address the high potential value pain points they often face.

# Essence Checklists Helping With Real Practitioner Pain Points

Recall the point made at the start of this chapter with respect to practitioners being skeptics when it comes to process improvement initiatives. They don't want extra work unless it helps them with the real difficulties they face.

One of those real difficulties practitioners often face relates to stakeholders who fail to provide information that software developers need to get their job done.

In the procurement case study discussed earlier in the book the root problem of the organization missing its schedule commitments was traced to the procurement department not getting the hardware ordered on time. This affected the software developers because it cut into their software test time. This caused real pain to the software developers.

Notice the definition of Stakeholders which is highlighted by the large arrow in Figure 13-2. Stakeholders are people, groups or organizations who affect, or are affected by a software system.

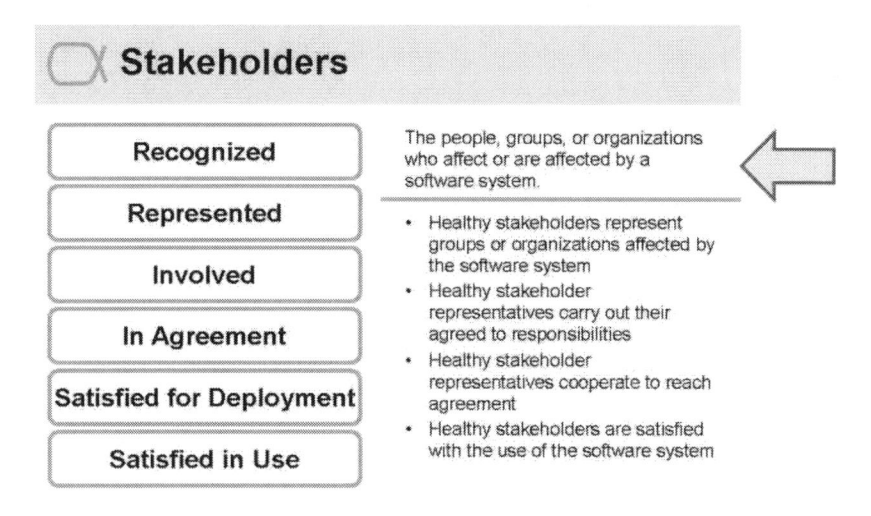

Figure 13-2 The Stakeholder Alpha

Some have argued that procuring hardware should not be considered part of a software engineering framework, but stakeholders are any one affected by *or who affects* the software system. Getting the hardware ordered on time affects the time available for the developers to test the software system. This is a common pattern that causes real pain to many software practitioners. This example demonstrates how the Essence framework isn't just a theoretical software engineering framework, but rather is focused on giving practitioners what they need to be successful with their software endeavors.

## How Esssence Checklists Differ From Traditional Checklists

Traditional checklists are what I refer to as "existence checklists." Examples include:

- Do you have a plan?
- Do you have a design document?
- Did you conduct a peer review?

Existence checklists are easy to use because the answer is a simple yes or no that requires little discussion. But existence checklists can lead to a *checklist mentality*, and they don't help us with questions related to *how well* we are performing or *how much* of a certain activity we should be doing.

Existence checklists are what many quality organizations use today and they are common among organizations that use the CMMI model. One reason for this is because existence checklists are easy for external appraisers to use, or external quality audit personnel who are not intimately familiar with the project they are auditing.

Now don't get me wrong. The CMMI model has helped many organizations and I am not suggesting that Essence should replace the CMMI, but this is an area where the use of the Essence framework could power the CMMI helping the organization's performance side.

Now lets look at some actual examples that demonstrate how many of the Essence checklists go beyond existence checks helping your practitioners get to their real pain points that can pay back high value when solved.

# Examples How Essence Checklists Go Beyond Existence Checklists

In Figure 13-3 you see four Alphas for discussion including Requirements, Stakeholders, Team and Software System.

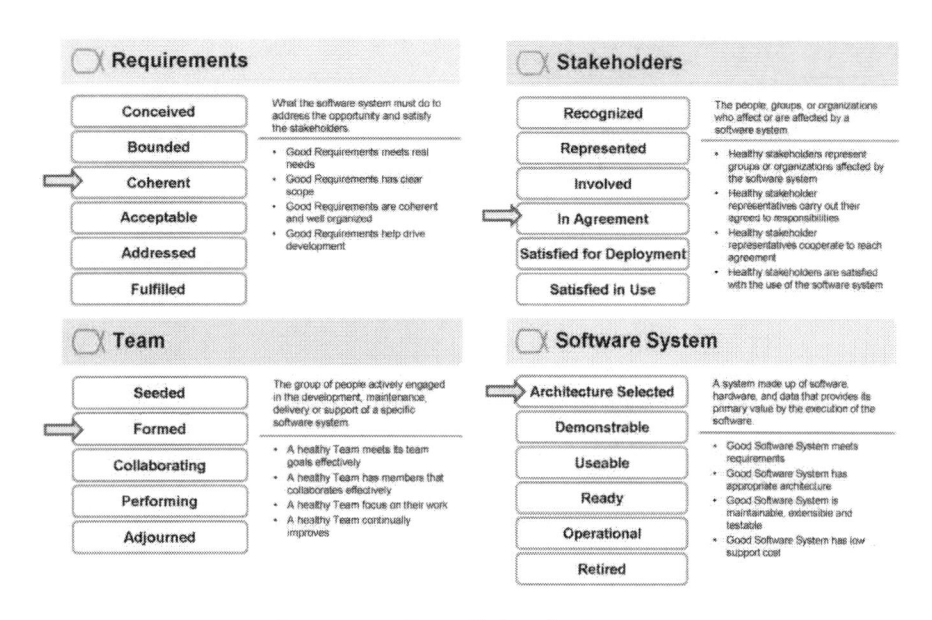

Figure 13-3 Four Alphas for Discussion

The four arrows in the referenced diagram indicate states of each of these Alphas I want to discuss to help explain how Essence checklists differ from many traditional checklists and how these checklists can help practitioners with their real pain points.

# Example 1: Checklist item Requirements Alpha Coherent State

A checklist item for the Requirements Alpha Coherent state is:

- Conflicts are identified and attended to

Note that just verifying that a requirements document exists won't be sufficient to verify that this checklist item is met. This checklist item is asking us whether or not we have conflicting requirements and if we do are they being addressed? Often the real pain that developers face originates from conflicting requirements that they don't know how to handle.

## Example 2: Checklist item Stakeholders Alpha In Agreement State

A checklist item for the Stakeholders Alpha In Agreement state is:

- The stakeholder representatives agree with how their different priorities and perspectives are being balanced to provide a clear direction for the team.

Note how carefully this checklist item is worded. It does not say all the stakeholders agree because on real projects this is often unrealistic. But while we may not be able to get all stakeholders to agree, we should be able to get their representatives to agree on how the priorities are being balanced to provide clear direction for the team.

We know that the root of many troubled projects and pain to many practitioners comes down to unclear direction for the team. This checklist item leads us to ask deeper questions related to the real health of the effort.

## Example 3: Checklist item Team Alpha Formed State

A checklist item for the Team Alpha Formed state is:

- Team members understand their responsibilities and how they align with their competencies.

Note how this checklist item leads us beyond just asking if a group of people have been identified. A team is more than just a group of people. This checklist item also reflects the reality that a formed team may have members with responsibilities that they may need help in carrying out. A healthy formed team doesn't mean a perfect team, but one where communication is open and honest about where competencies may be lacking so the team can take appropriate action to address those shortcomings.

# Example 4: Checklist item SW System Alpha Architecture Selected State

A checklist item for the Software System Alpha Architecture Selected State is:

- Architecture selected that addresses agreed to key technical risks

Note how this checklist item leads us beyond just asking if an Architecture exists. It leads the team to ask if they have agreement on the key technical risks, and if those risks have been taken into consideration in selecting the architecture.

This helps the team confront in an honest and open way the health of their proposed solution. This checklist item is found in the first state in the Software System Alpha. How often have projects found themselves in trouble because they have moved forward with multiple demonstrations of parts of the system to key stakeholders while leaving major technical risks looming to eventually cause great pain late in the project?

# Not Checklists an External Auditor Can Easily Apply

These checklists are not checklists an external auditor can easily apply by looking at a project from the outside. You need to be intimately involved in the project to answer honestly the questions that many of the Essence checklists lead you to ask. This is one reason why this framework is for practitioners and why practitioners need to be more actively involved when making key decisions that help to steer a successful endeavor. Recall that our framework vision called for placing the practitioner in charge of their own practices. It is the practitioner who is best qualified to apply the Essence framework because it is the practitioner who knows best how to answer many of the questions the framework will lead you to ask.

Traditional checklists lead us to simple yes/no answers that require little discussion. Many of the Essence checklists lead us to deeper questions, and deeper analysis– the kind that gets you to the real issues where practitioners most often need help.

## An Example of Using Essence as a Thinking Framework

During a review of the Essence specification in early 2013, one of the comments from a reviewer related to the fact that it was unrealistic to think you could get all the stakeholders to be involved and provide timely feedback, especially on large projects. There are two points to be made here. First, the checklists refer to stakeholder representatives because it is understood that not all stakeholders can be involved. Second, if a certain stakeholder representative is not providing timely feedback, then the team must decide how important this stakeholder representative is. The team could decide this is not a serious issue, or the team could decide to raise a risk. The right answer depends on the specific situation which the team understands best. This is an example of what we mean when we say the Essence framework is a thinking framework that helps the team make the best decisions given their situation.

Now lets talk about practices.

## Examples Essence Helping Detect Practitioner Practice Pain Points

Practices are not in the Essence Framework, but are defined as an extension. A practice in Essence is defined as:

*a repeatable approach to doing something with a specific purpose in mind.*

Any practice your organization may currently be using can be represented in the Essence framework. When you represent a practice in Essence it brings focus to at least one Alpha and at least one of its states.

If you can't find the Alpha your practice is helping, Essence leads you to ask:

*Why are we asking people to do this stuff (what is the value)?*

And if you find you have multiple practices all addressing the same state of the same Alpha, Essence leads you to ask:

*Why are we asking people to do this stuff (what is the added value)?*

Now think back to why practitioners are often skeptics of process improvement efforts. When your team answers these questions you may decide there are good reasons for what you are asking your practitioners to do. But you may also decide some of your practices can be made more lean helping your team focus their effort where the payback is most important now.

When you represent your practices in the Essence system you may also find you have gaps. That is, you may not have a practice that is helping to progress a certain Alpha to a certain state. When you locate these areas the Essence framework leads your team to ask:

*Do we need a better defined practice in this area?*

And

*Is this an area that is causing this team pain that needs to be addressed now?*

Using Essence helps a team bring the right issues to light by helping them ask the right questions leading to better decisions and more effective practices.

# Helping Us Understand Practitioner Frustration

The preceding discussion helps us understand at least one of the reasons why practitioners get frustrated and are often skeptical of process initiatives. While the CMMI is an improvement model that can also help to detect missing practices, its focus in the way many organizations have applied it has not been on helping to identify and improve overly heavyweight, redundant and inefficient practices which often cause great pain to practitioners. Recall our framework vision called for an aid that keeps our practices alive, lean and helping us when we need help the most, rather than burdening us with extra work that brings little or no added value.

# Example Essence Powering CMMI, Lean Six Sigma, Agile Retrospectives

Essence includes 6 essential competencies. Refer to Figure 13-4.

Figure 13-4 Essence 6 Essential Competencies

Competencies are the abilities needed. Each competency has five standard levels of achievement ranging from Level 1 Assists (ability to follow instructions and complete basic tasks) to Level 5 Innovates (a recognized expert).

As an example, the Stakeholder Representation competency is defined to be:

*A competency that encapsulates the ability to gather, communicate and balance the needs of other stakeholders and accurately represent their views.*

A common problem observed on many projects that can frustrate developers is when stakeholder representatives focus on just their own views, and fail to adequately represent the views of other stakeholders who they have been chosen to represent.

Competencies have been included in the kernel to help teams assess whether they have the right abilities, and the right level of abilities on their project. The CMMI, Lean Six Sigma, and Agile Retrospectives provide little help in this area. This is an area where Essence could power the CMMI, Lean Six Sigma and Agile Retrospectives helping you raise the competency of your people faster.

# Example of a Team Using Essence Independent of Practices

I would now like to give you an example scenario that demonstrates how a team could use the Essence framework even if they choose not to represent their practices in the Essence system. This example was developed by the SEMAT volunteers as part of the development of the Essence Users Guide.[50]

As the scenario begins a project is just kicking off with three assigned team members tasked to modernize a customer information legacy system for a retail hardware store chain. The team members decide to use the Essence framework in a planning poker style [57] to assess where they are.

The purpose of the scenario is to demonstrate the value of the Alphas, Alpha states and checklists. The team members walkthrough each of the seven Alphas independently assessing each and then they share their rationale with the other team members for why they think they have reached the state they selected and what is keeping them from getting to the next state.

---

[50] The Essence Users Guide is planned to be released late summer, 2014, and is intended to be used by practitioners, students, practice authors, educators and Essence tool vendors. I would like to acknowledge specifically the following SEMAT volunteers for their efforts on this scenario: Dr. Mira Kajko-Mattsson, Winifred Menezes, Barry Myburgh, Dr. Cecile Peraire, and Dr. Robert Palank. Refer to the SEMAT web site for more information (www.semat.org).

In the interest of brevity I will share the results of the team's assessment of just one of the Alphas–the Team Alpha.

After some discussion the team agrees that they have achieved the Team Seeded state, and their next target state is Team Formed state. Refer to Figure 13-5.

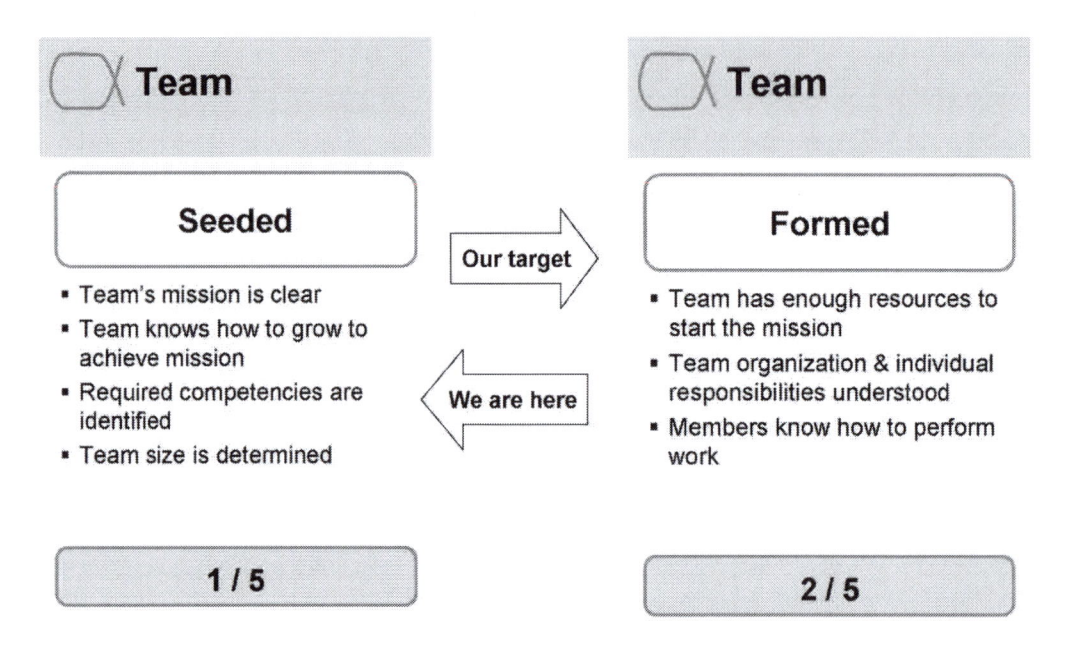

**Team — Seeded (1/5)**
- Team's mission is clear
- Team knows how to grow to achieve mission
- Required competencies are identified
- Team size is determined

Our target →
← We are here

**Team — Formed (2/5)**
- Team has enough resources to start the mission
- Team organization & individual responsibilities understood
- Members know how to perform work

Figure 13-5 Where we are, and next target state

The team agreed that it hasn't achieved Team Formed state because it hasn't achieved the following checklist item:

- *Individual responsibilities are understood*

What you learn from this scenario is how a team uses the checklists to help them reason about their current state and reach concurrence on the next actions that need to be handled to move forward on their project.

It is worth pointing out here that "*how much*" effort the team needs to put into achieving each Alpha checklist item is a decision that the team needs to make based on many factors including project specific issues, their own experience level, and the degree to which the team has worked together in the past.

It is the team members themselves who can best answer the "how much" question because it is the team that best understands their own situation.

What I find most interesting about this scenario– and most relevant to the challenge presented earlier in this chapter– is that when listening to the team member's discussion you hear about things that do not seem like big issues at all. In fact, they are things that seem so small it is easy to see how they could just be ignored. But these are also the types of small things that often when missed early lead to big things down stream.

# How Essence Relates to Two Types of Pain Points

With respect to the two types of pain points referred to earlier, Essence can help with type 1 in multiple ways including through the use of additional checklists and hints, similarly to what we saw earlier in the book in the procurement case study.

With respect to type 2 one thing I hope you are getting is that type 2 situations are best solved by the full team using the Essence framework as it is intended–as a thinking framework not to just reach a yes/no answer, but to find the right answer given the team's specific situation.

For the tough type 2 situations that seem to repeat at the worst possible time– just when you need to be on top of your game– Essence also provides patterns that can be defined as an extension to the framework similar to practices. One use of patterns is to remind practitioners of options and consequences of their decisions when faced with some of the tough common situations they face each day on the job.

The pattern construct in the Essence framework is a way to provide those reminders to practitioners of common situations they should be alert to, and possible options and consequences to potential related decisions that we discussed earlier in the book when presenting our thinking framework vision.

In the next chapter we take a closer look at Essence patterns and provide some example scenarios.

## Chapter Thirteen Summary Key Points

- It doesn't help to tell people to make sure they handle all the small things because not all small things lead to big things, and the real problem is that on the majority of projects there just isn't enough time to do everything.
- You can use the Essence framework whether you decide to bring your practices into the Essence system, or not.
- Existence checklists are easy to use because the answer is a simple yes or no. But existence checklists can lead to a *checklist mentality*, and they don't help us with questions related to *how well* we are performing or *how much* of a certain activity we should be doing.
- Essence checklists lead us to deeper questions, and deeper analysis- the kind that gets you to the real issues where practitioners most often need help.
- Using Essence helps a team bring the right issues to light by helping them ask the right questions leading to better decisions through better practice.
- "*How much*" effort the team needs to put into achieving each Alpha checklist item is a decision that the team needs to make based on many factors including project specific issues, their own experience level, and the degree to which the team has worked together in the past.

# Chapter Fourteen: Better Decisions, Better Practice, and Essence Patterns

*"Any intelligent fool can make things bigger, more complex, and more violent. It takes a touch of genius – and a lot of courage – to move in the opposite direction." Albert Einstein*

When do practitioners need help the most? If things are going well, the last thing a practitioner needs is extra work that doesn't add value. It is when things aren't going well and practitioners face a difficult decision– especially when they need to make it rapidly and often under stressful project conditions– where they need help the most.

We also know from experience that under such circumstances it isn't very useful to open a large complex process document. But if our practitioners have been recently exposed to the right common patterns with related options and consequences, there is greater likelihood of their making a better decision, even if it must be made rapidly.

## What is an Essence Pattern?

An Essence pattern is defined as:

- An arrangement of the other language elements (e.g. Alphas, Alpha states, competencies…) into a meaningful structure. [54]

Essence patterns can be used to define roles with responsibilities, and milestones. They can also be used as reminders to practitioners of common situations and related options and consequences. Essence patterns can be especially valuable to practitioners when they need to make a rapid decision.

It is the use of patterns as reminders to practitioners to help them make better decisions often under difficult situations that I want to explore in this chapter. Lets now look at a few examples.

# Pattern 1: The Dictated Schedule

Often the most difficult decisions practitioners face isn't "what" to do, but "how much" to do. We saw an example of this situation in the Dictated Schedule Scenario discussed in Chapter Ten. This scenario demonstrates one of the dangers in applying checklists with a simple yes/no "checklist mentality". In this example we examine the Dictated Schedule Scenario again, but now using the Essence thinking framework and the Essence pattern construct to show you how you can raise the visibility of common patterns in your organization leading to better on the job decisions.

**Pattern Start**

If you recall the Dictated Schedule scenario, the Project Leader when speaking to her team, said:

*"Management has dictated that we will not slip schedule, so everyone needs to do whatever it takes to get the job done."*

A developer responded:

*"OK. I am just going to get my code to run. I am going to skip my design review and I'll skip most of the testing I was going to do because management doesn't care if I follow the process."*

Let's now look at two Alphas relevant to this scenario— Work and Way of Working. Refer to Fig 14-1.

**Work**

- Initiated
- Prepared
- Started
- Under Control
- Concluded
- Closed

Activity involving mental or physical effort done in order to achieve a result.

- Healthy Work is sizeable, estimate-able and track-able
- Healthy Work breakdown reduces dependencies between work items
- Healthy Work management keeps risks, work and re-work under control

**Way-of-Working**

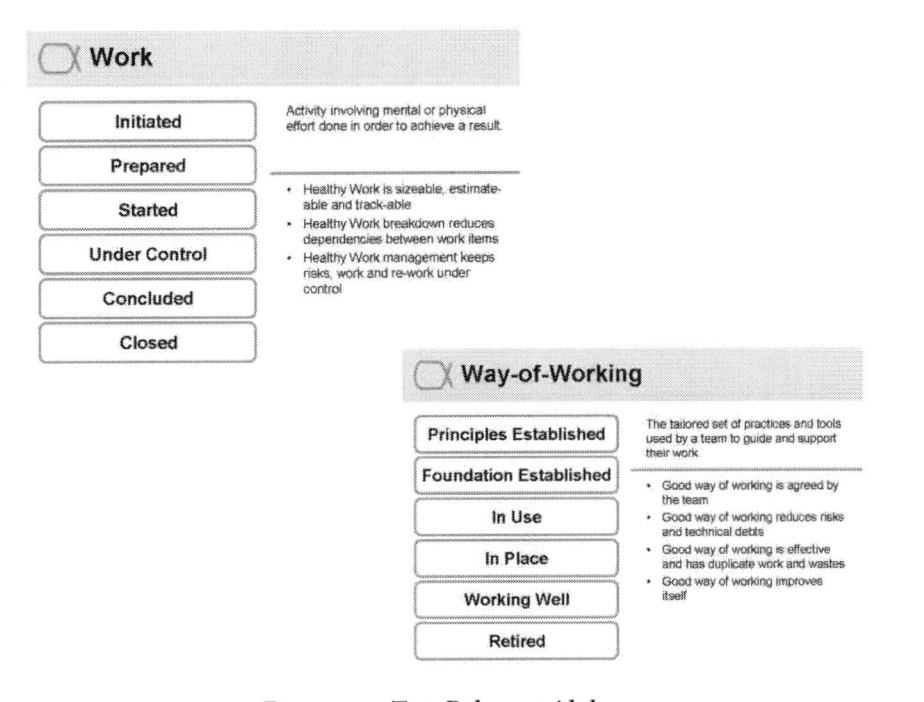

- Principles Established
- Foundation Established
- In Use
- In Place
- Working Well
- Retired

The tailored set of practices and tools used by a team to guide and support their work.

- Good way of working is agreed by the team
- Good way of working reduces risks and technical debts
- Good way of working is effective and has duplicate work and wastes
- Good way of working improves itself

**Figure 14-1 Two Relevant Alphas**

Let's assume in this scenario that the team had been using the Essence framework and had assessed their endeavor to have achieved the following states for these two Alphas before our developer made his decision: Way of Working Alpha *In Place* state, and Work Alpha *Under Control* state. Refer to Figure 14-2.

## Work — Under Control

- Work going well, risks being managed
- Unplanned work & re-work under control
- Work items completed within estimates
- Measures tracked

4 / 6

## Way of Working — In Place

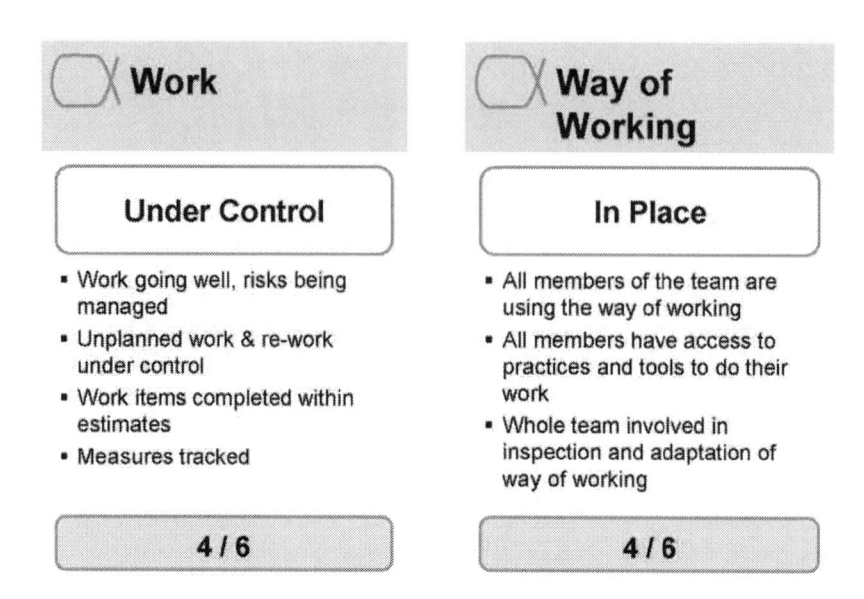

- All members of the team are using the way of working
- All members have access to practices and tools to do their work
- Whole team involved in inspection and adaptation of way of working

4 / 6

Figure 14-2 States Achieved

The achievement of these states was based on the team's assessment using the Alpha state checklists that indicated all of the team members were using the way of working that had been agreed to, the risks the team had identified were being effectively managed, and that rework was under control. Refer to Figure 14-3 for key checklist items that led the team to their current Alpha state assessment.

## Work

### Under Control

- Work going well, risks being managed
- Unplanned work & re-work under control
- Work items completed within estimates
- Measures tracked

**4 / 6**

## Way of Working

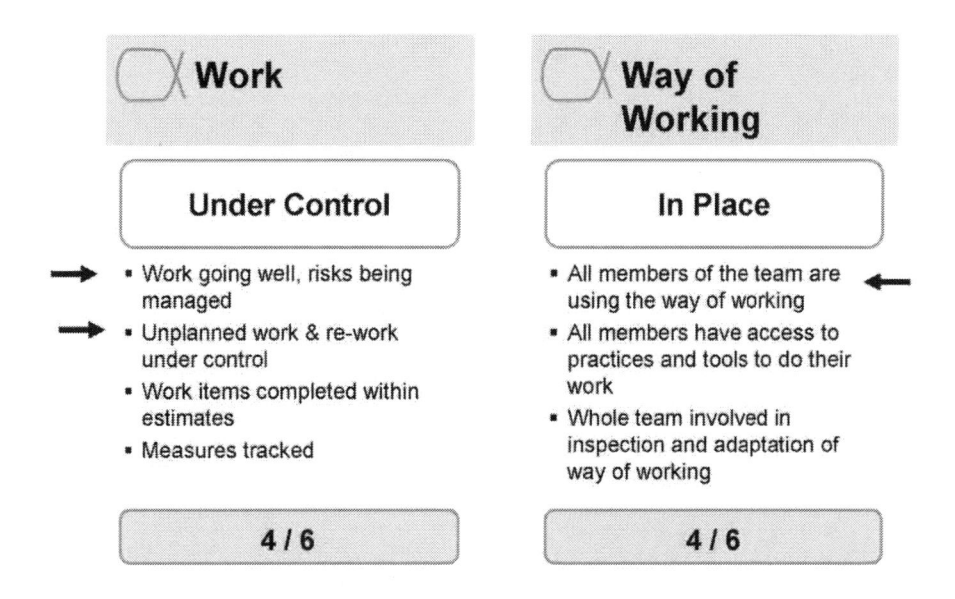

### In Place

- All members of the team are using the way of working
- All members have access to practices and tools to do their work
- Whole team involved in inspection and adaptation of way of working

**4 / 6**

Figure 14-3 Key Checklist Items Achieved

Lets also assume that the team's agreed to way of working included practices that required all architectural components to undergo a formal design review and multiple levels of testing.

Recall that in this scenario our developer made a decision not to follow the agreed to way of working because he saw no options to maintain the schedule that was being dictated and also follow the required practices.

Because of this decision the team updated their assessment of the state of their endeavor with respect to these two Alphas as follows: Way of Working *In Use*, and Work *Started*. Refer to Figure 14-4.

## Work — Started

- Development work has started
- Work progress is monitored
- Work broken down into actionable items with clear definition of done
- Team members are accepting and progressing work items

3 / 6

## Way of Working — In Use

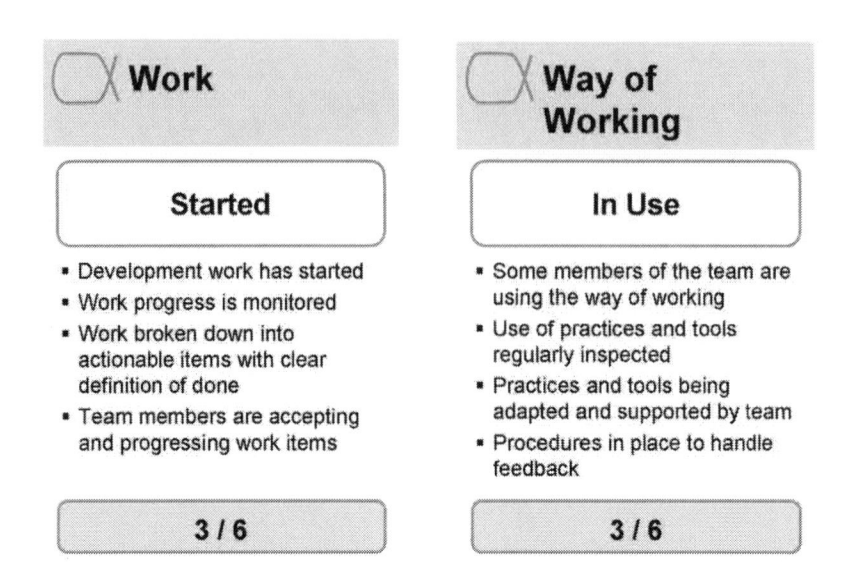

- Some members of the team are using the way of working
- Use of practices and tools regularly inspected
- Practices and tools being adapted and supported by team
- Procedures in place to handle feedback

3 / 6

Figure 14-4 Current States Fallen Back

This example demonstrates how a team can fall back.

The Way of Working had previously achieved the *In Place* state, but the team now assessed that they had fallen back to the *In Use* state because not all team members were following the agreed to way of working.

They also assessed that the Work Alpha was no longer in the *Under Control* state, but had fallen back to the *Started* state because the team recognized it now had a high risk going forward due to the lack of adequate testing and design reviews.

# Scenario Assessment and Consequence of Decision

Because of the way the team's design review and testing practices had been written the developer saw no options to comply with the agreed way of working and the dictated schedule by the project manager. As a result, he made a decision that caused the team progress and health assessment to fall back.

# A Decision that Could Have Improved the Team's Performance

So what could the team have done differently in this scenario that might have led to a better outcome?

One approach that might have led our developer to make a different and potentially better decision relates to the use of criteria in practices.

If this organization, instead of requiring that all architectural components undergo a formal design review and multiple levels of testing, provided criteria within their process description that allowed the developer to assess his or her situation and decide how much testing and how much design review was needed, a more practical and balanced solution might have been found.

Examples of the type of criteria that could have helped this practitioner make a better decision include considering:

- risk
- complexity
- effort required
- experience of the developer
- history of the component

Providing criteria to help practitioners make decisions related to "how much" testing and "how much" design and peer review are needed can be an effective way to aid practitioner decision-making. [23] By providing criteria in your processes you give your developers more options to find the right solution given their specific situation.[51] However, keep in mind that in order for team members to apply criteria as suggested in this scenario, they need to have more than just technical competencies. They also need an appropriate level of management competency to be able to effectively self-manage their work and their way of working.

For example, from the checklist items for the Work Alpha the following minimum essential competencies for self-management of work can be extracted: The ability to:

---

[51] In the Procurement Case in Chapter Seven we saw an example where the appropriate action could be captured through a small set of patterns with specific actions to take in each case. In many situations the appropriate action is not this straightforward. In these cases providing criteria to aid the practitioner's reasoning process may be more appropriate.

- Break work down into estimate-able work items
- Estimate work items
- Track work item progress
- Complete work items within estimates
- Identify and report risks.

## Scenario Summary

This scenario has demonstrated multiple key points to keep in mind when using the framework. First, teams can regress. You do not always move through the Alpha states in a strict linear fashion. That is, you can reach a state, and then fall back. Second, how you describe your processes (e.g. use of criteria) affects the degree to which practitioners can effectively address common situations faced with practical options to maintain desired team performance.

If criteria had been used in the case just described, the practitioner in our scenario could have potentially found an appropriate balance of testing and reviews that could have helped maintain the aggressive schedule being dictated by the project manager, minimize project risk and continue to follow the team's agreed way of working.

Keeping this scenario in mind can help developers make better decisions when facing similar situations on their own project. Refer to Figure 14-5 for a graphical reminder of the Dictated Schedule scenario using the Essence pattern construct.

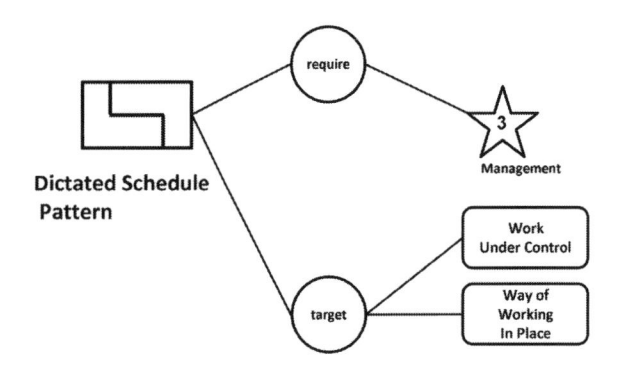

Figure 14-5 Two Dictated Schedule Pattern

The way a practitioner could use this diagram as a reminder is as follows. First, the practitioner is alerted that the pattern might be triggered whenever she senses a resource constraint that might mean she needs to make a "how much" decision. The diagram also reminds her that the goal is to maintain healthy states with the *Way of Working* Alpha in the *In place* state, and the *Work* Alpha in the *Under control* state. She is also reminded that in making her decision she will need to rely on certain management competencies (level 3).[52] [54] This could be a trigger to her to ask for help if she is uncertain as to whether she possesses the required competency that the situation calls for.

To be more explicit on the graphical reminder one could add the checklist items highlighted that still need to be achieved (or that we want to maintain). Refer to Figure 14-6.

---

[52] The diagram indicates that a level 3 of the Management Competency is required to apply this pattern. For more information on competency levels refer to Essence Specification submitted to the OMG.

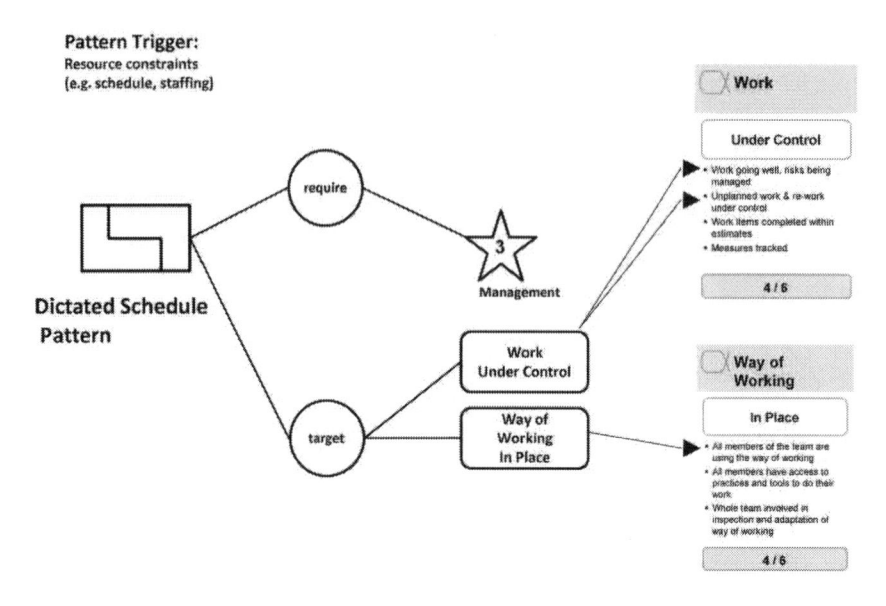

Figure 14-6 Dictated Schedule Pattern Target Checklists

Part of the power of these graphical reminders, and the use of the Essence framework in general, is that they give us explicit and visible objective evidence of a trouble spot, or a potential trouble spot. Too often when trouble is recognized only in an implicit way, things keep rolling along until the problem is much larger and you are forced to address it because you failed to detect it and take appropriate action at the right time.

As another option one could highlight more explicit information related to the competency needs in the scenario. Refer to Figure 14-7.

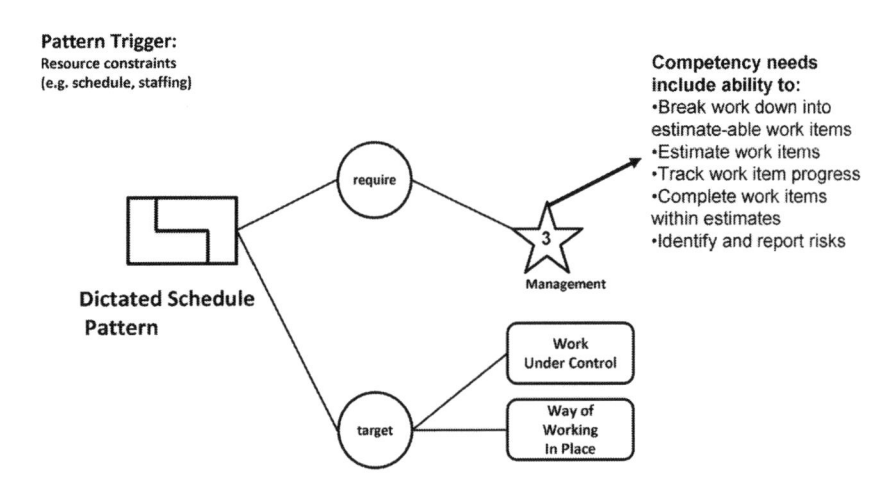

**Figure 14-7 Dictated Schedule Pattern Competency Needs**

Making competency needs more visible to your team could help to remind the team that less experienced team members may need some help, or they may need just a little coaching at just the right time. More on this subject in the next pattern, but before we move on I want to make one final point about this common pattern.

Some have argued that the pattern I described here is not realistic because requiring all software to go through a formal design review and multiple levels of testing is not something many organizations do today because of the change of thinking brought about by the agile movement. I disagree with this line of reasoning for the following reason.

While it is true that many organizations have lightened up on their formal processes and governance rules as a result of today's popular agile movement, it is also a fact that many organizations that develop applications with life-threatening consequences should their product fail, have continued to maintain rigid processes with strict review and test requirements. Furthermore, this may be perfectly appropriate given the risks in certain situations.

While developing processes with criteria can provide more options for your team, you need to consider your team's competency to apply criteria appropriately and the potential consequences of a wrong decision at the wrong time.

# Pattern 2: A Little Coaching at Just the Right Time

In the previous example we saw a common situation where higher self-management competency was needed for team members to make better decisions. When a team is continually making small changes to improve, team members may at times need a little extra help from their teammates, or a coach, to keep making better decisions. Following is a simple pattern to demonstrate this idea.

**Pattern Start**

I was coaching a team in an organization that was moving from a traditional waterfall development approach to a backlog-driven Scrum approach [58]. This organization had a strong history of projects being run by PMI certified project managers that required breaking down the work into detailed work packages, assigning specific tasks to specific personnel, and then monitoring the work through lengthy two hour weekly status meetings.

When I first explained in the training class to the development team members and the project manager the concept of self directed teams including how the team members sign up for tasks, and how the daily standup works through self-management, everyone liked the approach and seemed to understand it. But after the training ended the team had trouble following the new process they had learned. After the first few daily stand-up meetings the team members would go back to their workstations and just sit waiting for someone to tell them what to do next.

The project manager, who was also learning how to be a Scrum Master, came to me in private with a concern that the team wasn't getting it. There was no communication happening outside the daily standup. Issues were being raised in the daily standup, but they weren't getting resolved.

She asked me if she should initiate her usual two hour status meeting to drive issues to resolution. Apparently, this is what the team expected because that was what they had always been used to on their previous waterfall projects.

I replied:

*"No, that is not your role now. But you do need to coach the team right when you see an opportunity to remind them they must initiate meetings themselves when they*

*have an issue that needs resolution."*

She said she didn't think she would ever become a good Scrum Master and she expressed concern that her twenty years of experience as a traditional PMI certified project manager was now just a hindrance to her.

I told her that was not the case, and all her experience was still valuable and she should use it to sense when there is trouble on the project. The change that she would need to make in her own behavior as she moved from a traditional project manager to her new role as a Scrum Master was a change in style. I told her rather than direct the team members with tasks and drive issues to resolution, she now should take on more of a coaching perspective letting team members know it is up to each of them to work with other team members to achieve the project goals.

Sometimes, as we see in this example, when people first learn something new in the classroom they may indicate they understand, but when they return to their normal work environment they don't always sense the situations where their own behavior needs to change.

# Essence Thinking Framework Helping With Gentle Reminders

From an Essence thinking framework perspective the relevant Alpha in this situation is the *TEAM* Alpha. Refer to Figure 14-8.

# Chapter Fourteen: Better Decisions, Better Practice, and Essence Patterns

**Team**

- Seeded
- Formed
- Collaborating
- Performing
- Adjourned

The group of people actively engaged in the development, maintenance, delivery or support of a specific software system.

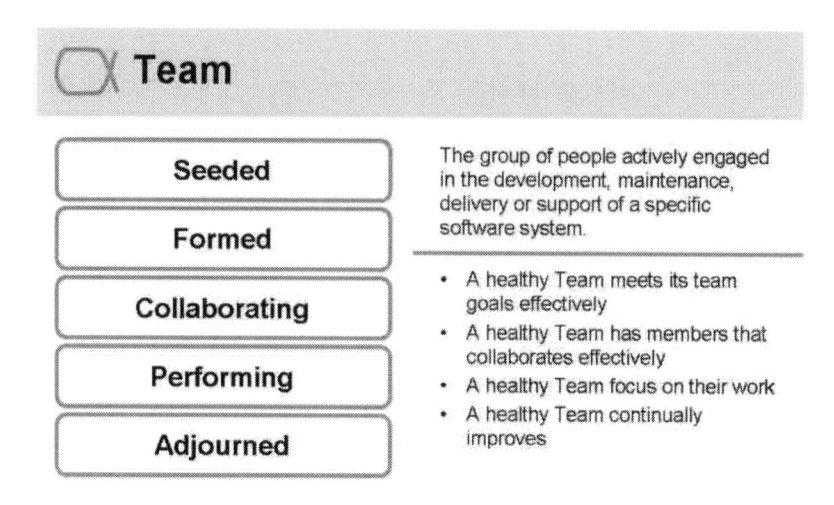

- A healthy Team meets its team goals effectively
- A healthy Team has members that collaborates effectively
- A healthy Team focus on their work
- A healthy Team continually improves

**Figure 14-8 Team Alpha**

The *Team* Alpha is defined as: *The group of people actively engaged in the development, maintenance, delivery or support of a specific software system.*

The Team Alpha has two states relevant to this situation: *Formed* and *Collaborating*. Refer to Figure 14-9.

## Team — Formed

- Team has enough resources to start the mission
- Team organization & individual responsibilities understood
- Members know how to perform work

**2 / 5**

## Team — Collaborating

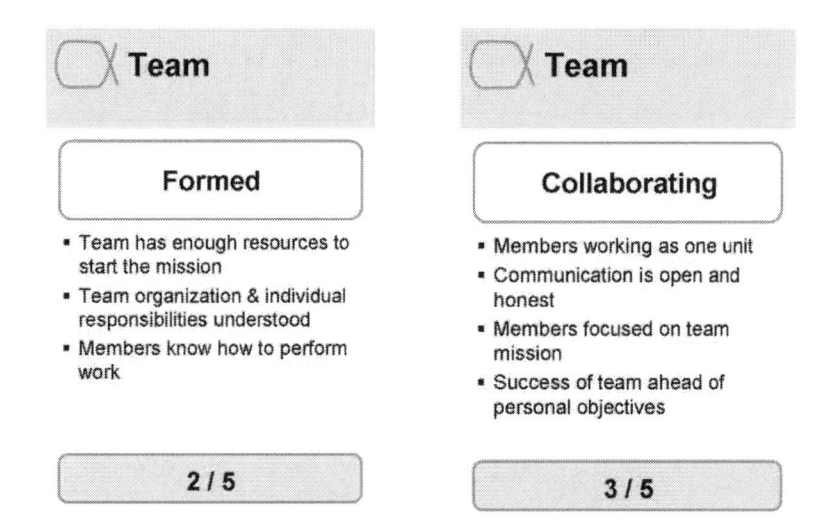

- Members working as one unit
- Communication is open and honest
- Members focused on team mission
- Success of team ahead of personal objectives

**3 / 5**

Figure 14-9 Team Alpha Formed and Collaborating States

The Team *Formed* state is defined as:

*the team has been populated with enough committed people to start the mission.*

The Team *Collaborating* state is defined as:

*the team members are working together as one unit.*

In this case the TEAM was formed, but they weren't yet collaborating. They didn't yet know how to work together as a team given the new rules of Scrum self-directed teams. In this case the team just needed a few gentle reminders at just the right time, and it didn't take long before they were off and running achieving the collaborating state.

Refer to Figure 14-10 for diagram of the "A Little Coaching at Just the Right Time" pattern.

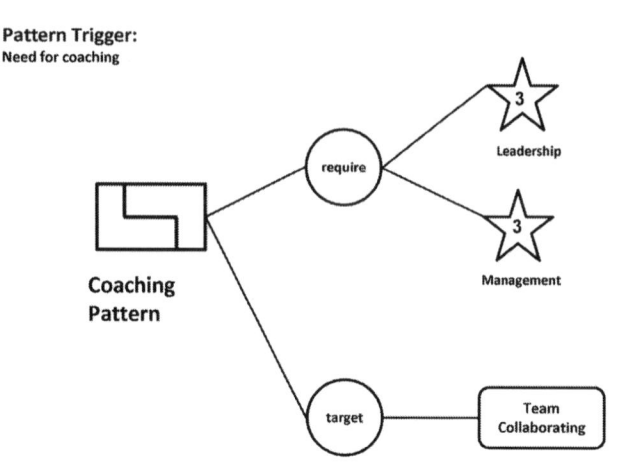

Figure 14-10 A Little Coaching at Just the Right Time Pattern

This pattern should be triggered when a coach or teammate senses that a team member needs just a little help right now.

# Apply the Right Practice at the Right Time With the Thinking Framework

What we have described in this simple example is not hard, but it takes practice at just the right time to be successful. I call this soft, or gentle reminders. Often just a little coaching is needed first– at just the right time– and after that your thinking framework can act as your coach reminding you– at just the right time– in the future [59]. One approach that has been used to help teams recall the Alphas and their states at just the right time is CARDS [55, 56]. You have already seen in this chapter a number of examples of the use of cards to aid progress and health assessments. Another way to use your cards is to add hints, or additional checklists, such as adding an additional checklist item to the Team Alpha collaborating state CARD to remind a newly formed team to call their own meetings when they have issues to resolve.

Refer to Figure 14-11 for an example of adding a checklist item to a state card.

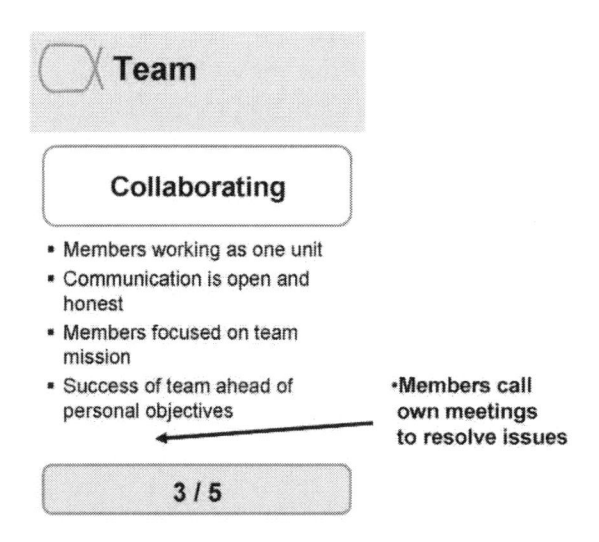

**Figure 14-11 Adding a Hint or Checklist item to a State Card**

I suggest just adding a hint in cases that may be temporary to help get a newly formed team going and add a new checklist item if you want to keep the reminder permanent. Once the team achieves the collaborating state it may not need to pay as much attention to this checklist item, but as new people come on board they may need to be reminded. Keep in mind that teams can fall back.

This was a simple example demonstrating how a pattern could be used to remind team members that they are each responsible to proactively resolve issues. In the following example you will learn about a common pattern that isn't as simple, nor as well understood, but can become critical to the ultimate success of your organization.

# Pattern 3: Keeping an Opportunity Alive and Healthy

In this example we look at the Opportunity Alpha. The goal of this example is to help you understand the importance of the opportunity, how the opportunity is typically progressed on an endeavor, and how it differs from requirements.

An opportunity is always essential to provide motivation for stakeholders, especially those who must fund an endeavor. Nevertheless, opportunities, once identified, often lose the attention of the development team. In this example we see a situation where the team knows the opportunity is in jeopardy, but only one team member knows what needs to be done to keep the opportunity progressing and healthy. This example should be of particular interest to project leaders.

Many teams don't think about the opportunity, but depending on your circumstances not constantly watching the opportunity could spell disaster to your endeavor regardless of the state of your requirements or the software system you are building.

**Pattern Start**

Before I started my consulting business I worked as a software developer with a software manager named Dave on a project I will refer to as Zack. The company had overpromised the customer to win the job and the schedule had been dictated by management without involving the team in estimating the real work. Zack was using an incremental approach and one of Dave's responsibilities was to run the daily build meeting and get the software ready to demonstrate to the customer for each increment.

Now Dave was an interesting character. It was not uncommon for him to be late to key project status meetings, or miss them altogether. He had been known to leave his office heading for a project status meeting, but then get stopped in the hall talking to a developer about a project requirement, and he would then forget where he was going and end up back in his office.

Another habit of Dave's that irked a lot of people on the Zack project was how he ran the daily build meetings. The Zack project had a well-defined process for running the daily build meeting that included a checklist of what was required for a software item to be acceptable for the next build. But at build meetings that were being held just before the end of an increment, Dave would often reject software that seemed to meet the checklist criteria, and then he would accept other software requests that failed many of the checklist items.

Given how Dave operated, it wasn't surprising that many viewed him as a poor manager, but when the customer came in for the incremental demonstrations they were always happy to see the functionality that was demonstrated.

Many on the Zack project thought Dave was just lucky and that the project would soon be cancelled. Nevertheless, it turned out that most of those people were dead wrong about Zack and about Dave. In the coming years Zack grew to be the largest and most successful project in the history of the company. Now lets look at a few of the relevant Alphas and as we do we will learn more about Dave and the Zack story.

The Opportunity Alpha is defined as:

*The set of circumstances that make it appropriate to develop or change a software system.* Refer to Figure 14-12.

**Opportunity**

- Identified
- Solution Needed
- Value Established
- Viable
- Addressed
- Benefit Accrued

The set of circumstances that makes it appropriate to develop or change a software system.

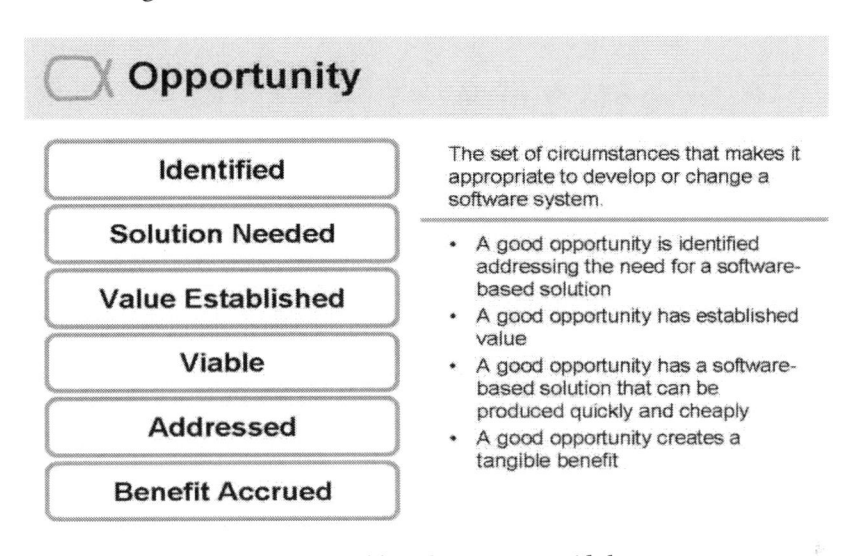

- A good opportunity is identified addressing the need for a software-based solution
- A good opportunity has established value
- A good opportunity has a software-based solution that can be produced quickly and cheaply
- A good opportunity creates a tangible benefit

Figure 14-12 Two Opportunity Alpha

One of the states of the Opportunity Alpha relevant to the Zack story is *Value Established*. Refer to Figure 14-13.

**Opportunity**

**Value Established**
- The value of a successful solution established
- Impact of solution on stakeholders understood
- Value of software system understood

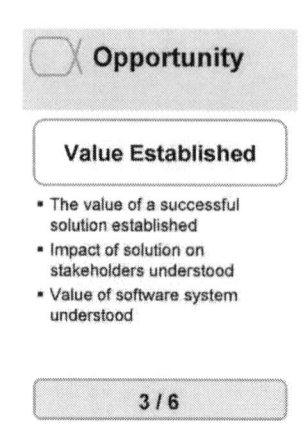

3 / 6

**Figure 14-13 Opportunity Alpha Value Established State**

*Value Established* state is defined as:

*The value of a successful solution has been established.*

Two of the relevant checklist items associated with this state are:

- *The value that the software system offers to the stakeholders that fund and use the software system is understood.*
- *The success criteria by which the deployment of the software system is to be judged are clear*

What Dave recognized early in the project was that the criteria Zack would be judged on was unclear, and he needed to get the right people to agree on that criteria.

Another relevant state is *Viable*. Refer to Figure 14-14.

**Opportunity**

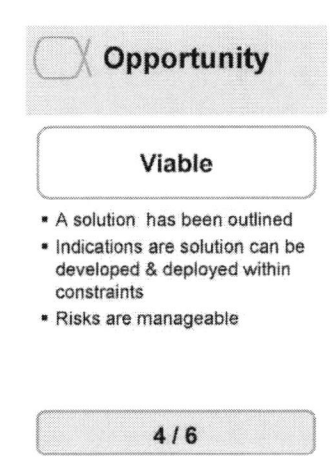

**Viable**
- A solution has been outlined
- Indications are solution can be developed & deployed within constraints
- Risks are manageable

4 / 6

**Figure 14-14 Opportunity Alpha Viable State**

The *Viable* state is defined as:

*It is agreed that a solution can be produced quickly and cheaply enough to successfully address the opportunity.*

A related checklist item is:

*It is clear that the pursuit of the opportunity is viable.*

Dave also knew he needed to show the key stakeholders that this opportunity was certainly viable. This leads us to another Alpha, the Software System Alpha and specifically its *Demonstrable* state.

The Software System Alpha is defined to be:

*A system that is made up of software, hardware and data that provides its primary value by the execution of the software.* Refer to Figure 14-15.

# Chapter Fourteen: Better Decisions, Better Practice, and Essence Patterns

**Software System**

States: Architecture Selected, Demonstrable, Useable, Ready, Operational, Retired

A system made up of software, hardware, and data that provides its primary value by the execution of the software.

- Good Software System meets requirements
- Good Software System has appropriate architecture
- Good Software System is maintainable, extensible and testable
- Good Software System has low support cost

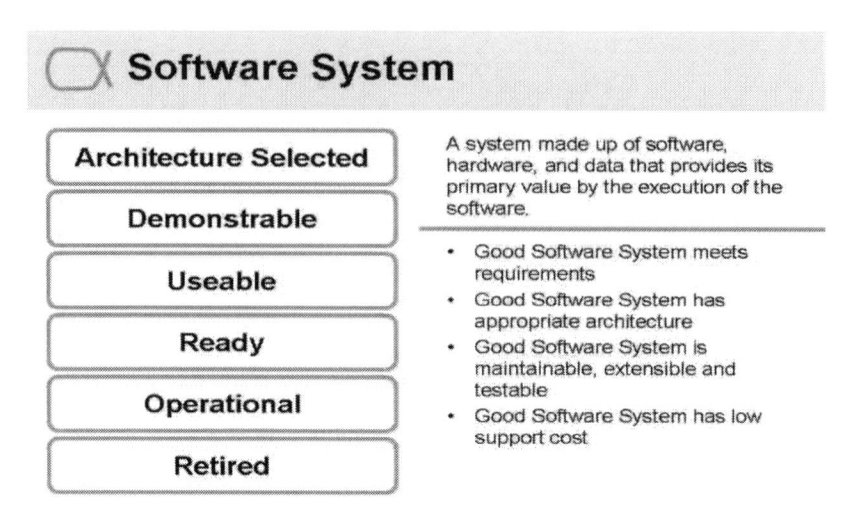

Figure 14-15 Software System Alpha

The *Demonstrable* state is defined as:

*An executable version of the system is available that demonstrates the architecture is fit for purpose and supports testing.*

Refer to Figure 14-16.

**Software System — Demonstrable**

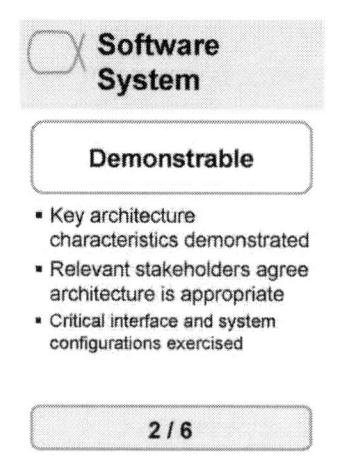

- Key architecture characteristics demonstrated
- Relevant stakeholders agree architecture is appropriate
- Critical interface and system configurations exercised

2 / 6

Figure 14-16 Software System Alpha Demonstrable State

The Software System had to be at least to a state where Dave could demonstrate to the stakeholders how it was fit for purpose. In other words, in assessing the progress and health of Zack, Dave knew he had not yet achieved the Opportunity Value Established and Viable States, but those were the next two states he needed to focus on and he knew to achieve those states he also needed to get the Software System to the Demonstrable state.

In Figure 14-17 you see Alpha states represented on Cards.

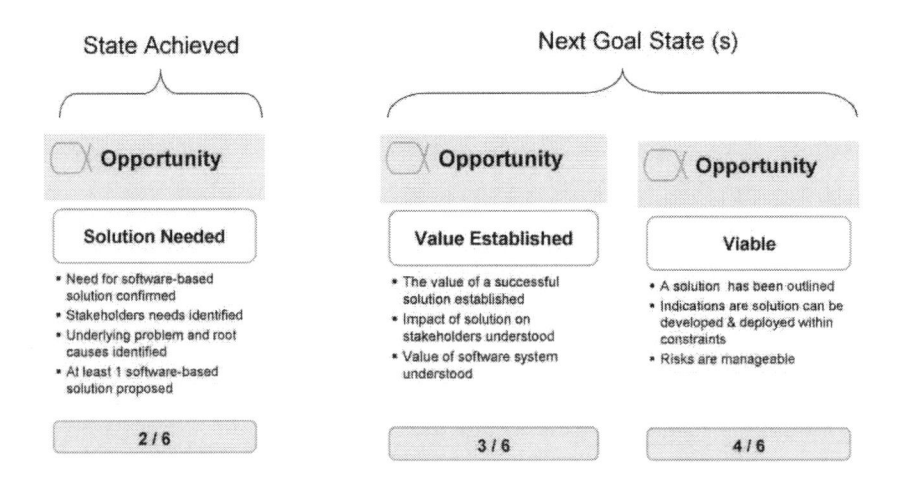

Figure 14-17 Progress and Health Assessment of Opportunity

Cards are a practical way to remember the framework. There are two types of Cards, Alpha Definition Cards and Alpha State Cards. The *Solution Needed* Alpha State Card on the left represents the state of the Opportunity Alpha that has been achieved. The two Alpha State cards on the right, *Value Established* and *Viable* represent the next goal states. These state cards provide a short-hand checklist visual aid reminder for each state shown. Earlier we saw an example of an Alpha Definition Card in Figure 14-1.

What few people knew, was what Dave was doing behind the scenes when he was missing Zack status meetings and what his thought process was when he was making *key decisions* on what to include in each build leading up to the next incremental demonstration to key customer stakeholders. He was spending much of his time talking to those key customers understanding what they really needed. He was learning what the key problem was that the key stakeholders that funded the system needed to be resolved.

*As it turned out there were many requirements that had found their way into the specification that really weren't that important to the key stakeholders who were funding the system and who would be the decisions makers to continue to fund the project.*

Dave learned from his discussions that the actual requirements had grown far larger than what the real opportunity required– that is the real need of the key stakeholders. Refer to Figure 14-18.

**Opportunity**

**Value Established**
- The value of a successful solution established
- Impact of solution on stakeholders understood
- Value of software system understood

3 / 6

Misaligned

**Requirements**

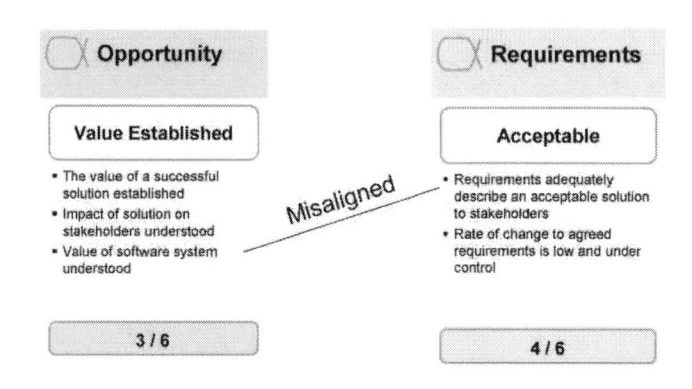

**Acceptable**
- Requirements adequately describe an acceptable solution to stakeholders
- Rate of change to agreed requirements is low and under control

4 / 6

**Figure 14-18 Misaligned Opportunity and Requirements**

He also knew from his experience if he tried to work the requirements without working the opportunity, the project would not survive. This example shows how projects can get in trouble when they focus on requirements without paying appropriate attention to the opportunity.

Dave figured out what it would take to save this project. He was constantly working to keep the opportunity alive by gathering the most current and relevant data to help him make the best decision he could each day on the job. Not everyone could have done what Dave did on the Zack project. It required at a minimum a *Master* level (level 3) of the *Leadership* and *Stakeholder Representation* competencies.

The *Leadership* competency is defined as:

*A competency that enables a person to inspire and motivate a group of people to achieve a successful conclusion to their work and to meet their objectives.*

When Dave was missing some of those meetings he was exhibiting leadership by taking the time to guide developers helping them focus on the most important requirements and capabilities of the software system.

The Stakeholder Representation competency is defined as:

*A competency that encapsulates the ability to gather, communicate, and balance the needs of other stakeholders, and accurately represent their views.*

The Master level 3 is defined as:

*someone who can apply the concepts in most contexts and has the experience to work without supervision.*

To keep the Zack project alive Dave needed to focus a great deal of his attention on key stakeholders ensuring the real needs were being met, especially the needs of those who were funding the system. He also needed to be talking to other stakeholders as well ensuring they understood the challenge the Zack project faced and why certain priority decisions were being made.

Refer to Figure 14-19 for a diagram using the Essence language pattern construct to graphically abstract key characteristics of this scenario.

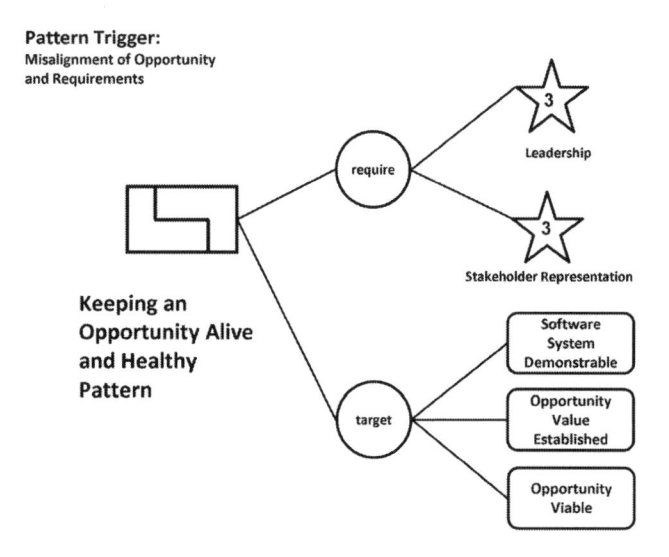

Figure 14-19 Keeping an Opportunity Alive and Healthy

This diagram highlights the fact that the trigger criteria for this pattern is a misalignment of the opportunity and requirements, and this pattern *targets* the

Software System *Demonstrable* state, the Opportunity *Value Established* state, and the Opportunity *Viable* State, and it highlights the fact that to *perform* this scenario *requires* the level 3 *Leadership* competency and the level 3 *Stakeholder Representation* competency.

Recalling this pattern could trigger a practitioner to recognize the lack of adequate leadership or stakeholder representation competencies on an endeavor where the opportunity and requirements are misaligned. This recognition could lead a practitioner to raise the need for action on their project if a similar situation is observed. It could also trigger the recognition that a project's opportunity might be in trouble and that action needs to be taken to raise the priority of getting the project to the Software System *Demonstrable* state and/or the Opportunity *Value Established* and/or *Viable* states.

As another option one could enhance the diagram with the target checklist items that are causing the project trouble and require special attention. Refer to Figure 14-20.

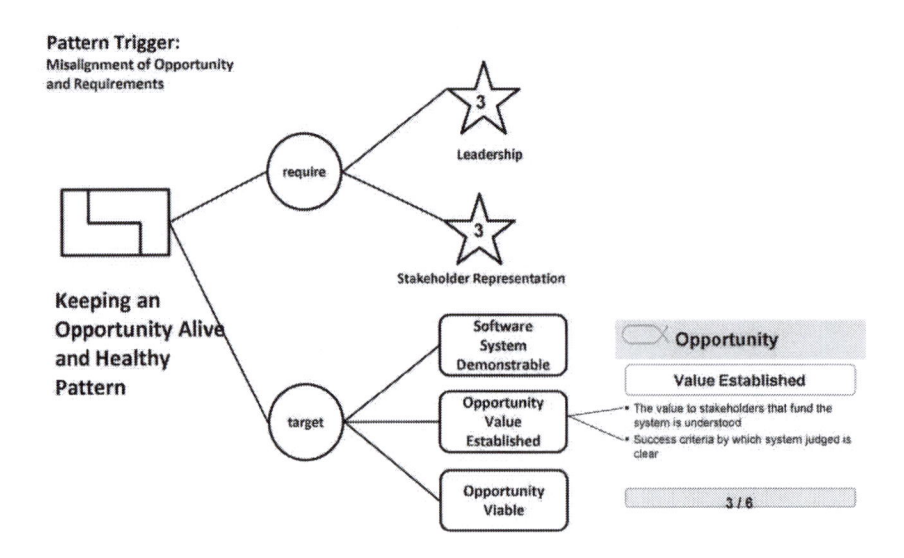

Figure 14-20 Pattern Diagram Enhanced with Target Checklist Items

This scenario demonstrate a key discriminator of the Essence approach which is how the framework can help practitioners see more clearly where they need to be placing their focus to achieve successful results.

Pattern diagrams can be used to support training, or they could be used by practitioners during an endeavor to help them recall common situations and related options and consequences helping them make better and more timely decisions. Other options practitioners have when using the framework to aid decisions include adding additional checklists, or hints (examples provided in appendices), or more specific information related to competency needs.

## Four Key Points from the Zack Example

There are four key points you should take from the Zack Example.

*1. Keeping an eye on all the Alphas:*

First, someone needs to be keeping an eye on, and thinking about, each of the Alphas, such as the Opportunity. I have watched great potential projects get cancelled because organizations stopped watching and working the opportunity at the wrong time.

*2. Effort will vary:*

Second, you may not need to spend as much time on your opportunity as Dave did. The effort required for each Alpha will vary dependent on the circumstances of your endeavor. The Alphas remind us of the essentials that someone needs to keep an eye on, but they don't tell you how much effort will be required for each.

You may be able to do a quick assessment and then put very little additional effort in because you find out the opportunity is being worked by someone else in your organization, or there is very little risk to the opportunity due to your specific project circumstances. Nevertheless, everyone involved in software development should understand why working the opportunity is essential on all software endeavors so they can be aware and communicate a key risk if they see it on an endeavor they are involved in.

*3. Reminders, not answers:*

Third, the Essence thinking framework did not tell Dave how to solve his opportunity challenge. The Alphas don't give you answers[53], but they do help you ask the right questions leading to better decisions and a better chance of project success.

---

[53]While Alphas don't give you answers, patterns could give you answers depending on how much detail you decide to provide in your patterns. Refer to the Procurement Case (Chapter Seven) for an example of patterns captured in a table format that provides answers in the form of actions to take.

*4. Practice every day*:

Fourth, Dave was practicing his opportunity practice every day on the Zack project because he assessed it to be the most important practice to keep the project alive. He was refining his opportunity practice and improving his own performance each day as he learned more about what Zack needed to be successful. This is what we mean by saying your practices come alive with the Essence approach. They are what you do each day, not just what someone thinks you should do.

# Summarizing the Idea of Essence Patterns

I have throughout this book through stories and examples explained the power that patterns can bring to help both individuals and organizations improve and maintain higher performance. In this chapter I provided just a few examples demonstrating how the Essence Framework (Alphas, states, and checklists) together with the Essence pattern construct could be used to help practitioners recall common situations and related options and consequences leading to better decisions when facing similar situations–especially those situations practitioners often face under today's common workforce pressures. Additional pattern examples focusing on other Alphas are provided in the Appendices to this book.

# Chapter Fourteen Summary Key Points

- If things are going well, the last thing a practitioner needs is extra work that doesn't add value.
- Essence patterns can be used as reminders to practitioners of common situations and related options and consequences. They are especially valuable where practitioners may need to make a rapid decision under a stressful situation.
- Providing criteria to help practitioners make decisions related to "how much" can be an effective way to aid practitioner decision-making. But you must consider the potential risks given your own situation.
- In order for team members to apply criteria as suggested, they need to have more than just technical competencies. They also need an appropriate level of management competency to be able to effectively self-manage their work and their way of working.
- Teams can regress. You do not always move through the Alpha states in a strict linear fashion.
- Often just a little coaching is needed first– at just the right time– and after that your thinking framework can act as your coach reminding you– at just the right time– in the future.

# Chapter Fifteen: Conclusion

A few years after I started my consulting business I went on a trip with my wife to visit our daughter who was studying music in Boston. It was a beautiful fall day and we spent much of it walking outdoors. Toward the end of the day I was getting concerned that our visit was keeping her from important practice. When I mentioned it she brushed the comment aside saying:

*"A lot of the other students at the conservatory practice their instrument a great deal more than I do. They will spend sun up to sun down practicing and some teachers expect that."*

But she also said that didn't work best for her, and she had actually figured out a better way to improve her flute performance. She said she could practice playing the flute without actually playing it. For example, while she is standing in line at the supermarket just waiting she thinks in her mind what it feels like to play a certain sequence that she's been working on and having a little trouble with. She said by getting her mind to think about how it feels to play the hard part her fingers start to think about it too even though there is no flute there. While explaining this process she moved her fingers toward her mouth holding them on each side as if playing an imaginary flute.

I have throughout this book stressed the importance of focusing on your repeating specific weaknesses. It is your repeating specific weaknesses that stand between where you are and the sustainable higher performance you desire. You can take control of your repeating specific weaknesses through the quality of your decisions. The quality of your decisions is driven by the quality of your practice. The quality of your practice is driven by what you focus on each day when you are preparing for your upcoming activities. Practice at the right time focused on the right things is far more important than the amount of time you spend on it. Done right, your practice can be short in duration, but it needs to be done consistently and with your undivided attention. When this form of improvement is placed on hold the impact to you and your organization's performance is often far greater than most realize.

I stated in the Introduction to this book that organizational process and performance improvements can be broadly grouped into two categories: Technology and major organizational changes looking to the future, and smaller changes that can help practitioners every day. One of my goals in this book has been to present the case that the speed of change we are all witnessing in today's world requires a rebalancing of how organizations view and prioritize their process and performance improvement initiatives across these two broad categories.

Let me now make a related observation. Organizations that use the CMMI high maturity practices or Lean Six Sigma techniques use statistical practices, such as hypothesis testing, to arrive at improvement decisions. This approach has been compared to the American judicial system in that we presume the individual charged with the crime (e.g. candidate root cause) is innocent until proven guilty [60].

With hypothesis testing we must be 95% confident based on the evidence before we convict. Or in the case of organizational improvement, we must be 95% confident based on the evidence before we make a decision to change and improve. In other words, the decision to improve or not to improve comes down to a discrete yes or no answer based on objective evidence.

This method has worked well for many organizations because it has helped them build a case that they can take to senior management or key stakeholders who they must convince to fund the improvement. While this approach continues to make sense today for major technology changes looking to the future, many of the potential small changes practitioners could be making each day to aid their performance do not come packaged with simple yes/no answers.

If we want to achieve and sustain higher performance in today's ever-changing world, we need to be encouraging our practitioners to make the kind of small changes every day that can help them learn and grow.

For those who worry about the potential negative ramifications of moving more control of our processes to our practitioners and allowing change in the midst of an endeavor, keep in mind that one of the primary reasons we emphasize keeping each change small is because it is then followed with rapid feedback and another small change/correction mitigating potential negative effects. Furthermore, no one is better positioned to make the best decisions on these small changes than our practitioners whom we must trust to get the job done.

I also stated in the introduction to this book that today we need a culture shift in how performance improvement is viewed and implemented. For years I didn't play golf because when I thought about golf the patterns in the front of my mind left me with no options. My memory of practice was that it was hard, it only helped me get to a certain level of proficiency, and from there I didn't know how to keep getting better. This led to my no longer enjoying the game.

Geoff Colvin makes the point in his book that "*deliberate*" practice is hard and you have to do a lot of it to become great. But it has been my experience if you make yourself do too much of anything and you do it at the wrong time it can have a negative effect. It can keep you from improving and sustaining that improvement because to improve you must first want to improve, and second you must know how to improve.

Today I see too many people learn the fundamentals of a job at work or a sport they seem interested in and soon after think that's all there is to it. This leads to boredom and looking for something else without fully realizing how much better they could get. Once you learn the essence of achieving and sustaining high performance you will also learn that there is no end to how much better you can get. Learning how to get better at anything is about learning more about yourself, your own personal habits, your own strengths, your own weaknesses, and learning when and how to ask for help.

Today by using the techniques described in this short book I have changed the way I feel on the golf course, and I find there is no end to my potential golf game performance improvements. To get better and continue getting better it takes a never-ending life long process of honest assessment and communication, first with ourselves and then with others. Today for me the golf course is no longer a place of frustration, but rather a place that is ripe with options and continual opportunities to learn new things. It is one of the places I go to grow.

Unfortunately, I see the work environment in many organizations today similar to the way I viewed a golf course when I was young. I see organizations where the people appear to be working hard at improving, but they aren't getting the payback and so they are frustrated, and looking around wishing they were somewhere else. You can change the culture of your workplace. The first place to start is to change your own personal work habits.

Following are what managers can do to help their organizations achieve and sustain their performance goal:

- Encourage sustainment sessions as follow up to training where people can ask specific questions about their tough challenges currently being faced on projects
- Use coaches to provide more on the job assistance
- Use scenario training (patterns) where the scenarios are tailored to the high value common issues being faced in your organization
- Let your people know it is ok to stretch themselves on the job taking small risks to perform better
- Authorize your people to locate and solve high value problems, not just collect data.

For individuals looking to improve their own performance:

- Block out time on your calendar to help yourself practice and be prepared before important endeavors such as status meetings with Senior Management, or speaking to a co-worker (this does not need to take a lot of time)
- Keep notes related to your own experiences and refer back to them frequently
- Build your own structured real stories and scenarios that you can review at just the right time to help you build the positive and accurate patterns you can quickly access right when you need them

In the introduction to this book I asked the questions:

*"Do you think you need to rehearse your practices to gain the potential high value payback? If you believe that preparing is important before you perform, then how should you go about doing it, and when should you do it?"*

I have tried to make the case throughout this book that to gain the potential high value payback you need to be practicing (or rehearsing) to be ready to handle the common difficult situations that arise that tend to hurt our performance the most. Only you know what those situations are and only you know when they are most likely to occur. The time to rehearse (or prepare) is close in time to when that situation is most likely going to happen and you should prepare by considering the specifics of the situation and the options you face. This will in turn help you to make the best possible decision maximizing the likelihood of achieving a high value payback for your efforts.

With respect to my comments on the need for a culture shift, we need to move away from the view of process improvement as a separate and distinct activity from project execution. I am not suggesting that all distinct process improvement activities should go away, only that a *rebalancing* occur between the distinct process improvement activities related to technology changes for the future and the activities which aid our workers today. High value sustainable performance in organizations of any size is achieved one person at a time and must extend beyond the technology side.

For years I have shared the simple techniques and practice aids discussed in this book through my workshops and I have had participants come back and tell me how they have helped them. But just learning about these techniques in a workshop will not help you achieve and sustain higher performance.

To get better at whatever you want to get better at requires more than training. It requires continual practice and not just any kind of practice, and certainly not the kind I did when I was young. I am talking about a very specific form of practice that is focused on your specific domain, your specific weaknesses, and the specific environment in which you must perform each day. You have the roadmap. The rest is up to you.

---

# Epilogue

In the Introduction to this book I stated that today many are turned off and have tuned out when it comes to the multitude of process and performance improvement approaches. I also stated that we have taken a simple idea and made it far too complex and in so doing have lost the spirit and intent of performance improvement.

Throughout the book I used a personal improvement story to help the reader think outside the box about performance improvement. In this epilogue I provide another sports analogy to demonstrate how these practices could be applied in a different domain.

When I was about thirty years old I started running to keep myself in shape. To stay motivated I also began participating in local road races. In the 1980's I ran three marathons, but the year after each I found myself suffering from a host of injuries accompanied by recurring knee, groin, lower back, and leg pains. After the 1988 Marine Corp Marathon in Washington, D.C. my recurring injuries kept me out of running for most of the following year. An orthopedic doctor in my home town diagnosed my problem based on what he saw on my MRI (MRIs were fairly new in 1989) as "probable Avascular Necrosis" and told me I would need a hip replacement.

I decided to seek a second opinion from a specialist in Boston who told me the spot on the MRI my home town doctor had referenced with his diagnosis was caused by a failure to properly set up and align the MRI machine. He also said he didn't know what was causing my pain, but that I should start running again and run through the pain as long as it didn't get a whole lot worse. He said, either you'll get better or you'll break it. "If you break it," he said with a slight smile, "come back, and then we'll know how to fix it."

So I went out and ran and it hurt. But the next day it didn't hurt a whole lot more so I ran a little further. The following day it still hurt but still the pain didn't seem to be a whole lot worse, and so I ran a little harder. I kept running just a little more each day, and one day a few months later I realized the pain was gone. After that experience I kept running, but I didn't run another marathon for 15 years because

I was afraid I would aggravate my trouble spots. However, during those 15 years I made a number of small changes to the way I run to try to help avoid the likelihood of my recurring trouble spots. I will explain those changes in a moment.

When my daughter was going to college in Boston fifteen years later I decided I wanted to see if I could qualify for and run the Boston Marathon. I trained for and ran the Scranton Steamtown Marathon in 2004 to qualify and I ran the Boston Marathon in 2005, 2006 and 2007. The day before the 2008 Boston Marathon I had a freak injury to my right calf. I thought it was a muscle pull, but it turned out to be a bad cramp. I was back to running within a few days, but I had to miss the Boston Marathon that year.

A few weeks after my problem at the 2008 Boston Marathon I asked my wife to go with me to the Vermont City Marathon. I felt that I needed to run a Marathon since I had trained hard for Boston and was unable to compete that year. My first mile of the Vermont City Marathon was slow, but I felt good and quickly picked up the pace. Before I knew it I was running 7 minute and 45 second miles which is too fast for me in a marathon. At the twelve mile mark the same calf that gave me trouble the day before the Boston Marathon locked up. "Oh no" I thought, "I've still got 14 miles to go. I'm cooked!"

But then I recalled something I had read in the book "*Again To Carthage*" [61], the long-awaited sequel to the classic, "*Once a Runner*"[62], by John L. Parker. In the marathon scene Quenton Cassidy gets a cramp in his side but is able to literally run it out of his body. I didn't know if this was actually possible. In the past whenever I had a bad cramp or muscle pull I would stop running and use ice and rest to fix the problem. But now I was thinking about how fast I had recovered from what felt like a similar problem just a few weeks earlier in Boston. There didn't seem to be any lingering negative effects from that experience. If this was just a cramp and not a muscle pull, then what was the risk in just continuing and seeing if I could run it out of my body? If I quit and sought medical help a massage therapist would use force with their hands to try to drive the cramp out. So I was thinking, why can't I just drive it out myself by running through it? I decided to just slow down a bit and see how far I could go. I've got 14 miles left, I thought to myself. I'll just take it a mile at a time and reassess how I feel each mile.

For the next few miles it wasn't getting any better, and maybe a little worse, but not a whole lot worse. With 12 miles to go my right calf was still very tight, and it didn't seem to be getting better. Of course, the prevailing wisdom would be to just stop running. It's not like I'm trying to qualify for the Olympics. A fundamental rule of running says when something hurts, stop, apply ice and rest. Otherwise you might do serious damage. This is a traditional rule all serious runners know. And I know there is truth to this rule. But the fact is that I do break many of the traditional rules of good running.

I didn't do this when I started running. The rules of running I break today are rules I have modified based on years of experience learning *specific* things about my unique body, and how it reacts and feels to me in different situations. Now let me explain some of the changes I have made in my running, and how my modified rules of running are based on my unique repeating specific weaknesses.

I used to frequently get injured when I ran too much. But over the past 20 years I have learned to run for the most part injury free. While I do follow rules they aren't the fundamental running rules you read about in popular books, or rules you would expect. In fact, with the rules I follow you would think I should have more injuries, not less. Most books about running will tell you to rest regularly. They tell you that you need rest to let your body recuperate.

I used to tell people I only run six days a week. I always take one day off to rest. I used to tell people that because I used to do it, and I used to do it because I thought that was what I needed to do to keep from getting injured. But the fact is I sometimes run for months without a day off. In fact, it is not uncommon for me to run an entire year without taking a day off.

So how do I stay injury free? I use some other techniques that I have figured out work for me because they have been developed based on my unique specific conditions. First, I know my body and I am constantly aware of and know how to interpret the signals my body is sending me. I do stop running at times but those times are dictated by my interpretation of the signals from my body, not the calendar telling me its Thursday.

I stop or slow when things don't feel right, but this doesn't mean I stop running every time I have a pain in my knee, or leg or hip. I have learned to distinguish different types of pain in different parts of my body. There are many types of pain that one can and should run through, and there are other types that should tell you to stop immediately. I know what specific pains feel like that I can run through, and I know how to sense the kind that is signaling me to stop.

I used to just stop every time anything hurt, but you can't get better if you stop every time something hurts. If you want to get better you have to learn how much and when to push, and how much and when to back off.

Unfortunately I can't give you the exact rules that are best for you because the answer is different for each person and organization. If you want to get better and keep getting better you have to figure out the rules that are best for you, practice continually, and realize the rules you follow today will change tomorrow. Now let me tell you more about my rules and why they work for me.

I now do almost all my training indoors on a treadmill. I used to run exclusively outdoors on the road. This means I am always at home when I am running (or in a hotel). I have created an environment that is conducive to taking the right action, when I need to take that action—like stopping immediately when I get a signal that says something is wrong and it is the kind of signal that tells me to stop running RIGHT NOW. I also have eliminated a number of potential areas that could cause injuries and have in the past. By running on a treadmill I don't get twisted ankles or pulled muscles from stepping in potholes, or slipping on ice, or tripping over street curbs. I don't get stress fractures from "pounding the pavement." Running on cement is hard on your body. Treadmills reduce muscle tear down. All of these things I have learned through my experiences of running for many years.

The Boston Marathon is always on Patriots Day, the third Monday in April. Temperatures can reach the seventies in late April in Boston. Training for the Boston Marathon when you live in the cold and snowy Northeast is not conducive to preparing for the likely conditions on race day in Boston. By running on my treadmill indoors at seventy degrees I acclimate my body to the potential race day conditions. I find it interesting that most running books I have read don't talk about these other *options* to effectively manage your running.

If you read books or articles that claim they can help you get better you'll find there are two types. There are books about fundamentals. These are useful, but will only get you so far. The other type is about individuals or teams or organizations that have proven to be high performers. In these books specific secrets to success are shared. But what these books or articles don't tell you is that if you want to reap the benefits of the secrets to success you'll need to first discover secrets hidden inside yourself.

I now know many things about myself that I didn't know when I first started running. My body is not perfect. I have curvature of my lower spine. This I have discovered is a source of my recurring knee, lower back and hip pain. To avoid these common repeating ailments I now run with orthotics and a lift in one shoe to compensate for the curvature in my lower spine. I have also changed my running style. I now run with more flex in my knees and I tilt forward (rather than running upright) to take stress off my pelvis and back. I *feel* more like a *"spring"* as I run. This style takes pressure off specific spots in my body where my past injuries have most frequently occurred.

As I get older I hear more and more from my running friends that they can't run anymore. Some of them have given up running (and taken up golf) because they believe it's easier on their body. They tell me their knees can't take the pounding anymore.

I trained again for the 2009 Boston Marathon and completed it successfully. When training for the 2010 Boston Marathon at 60 years old I found it difficult to build up my miles as I had done in the past. I recall thinking how much more painful my long runs were becoming. My quads hurt, my knees hurt, my Achilles hurt and my lower back which hadn't hurt in years was waking me up at night. I know I will not be able to run forever.

I was thinking maybe I should just go back to the shorter distances, and give up the marathon again. But before I did that I decided to try one more thing first. I started taking notes each day just writing down how I felt. I used to do this every day, but I discovered I had allowed myself to fall out of the habit.

After a few weeks of daily note taking, I stepped back and read them. It was then I realized some things I hadn't noticed before. It struck me that I was still running at the same pace on the treadmill for my long runs that I had been running for the last five years preparing for Boston. And it struck me that I still was not taking a day off for weeks at a time. So I decided to try a few *small changes* to see if it might make a difference. First, I started forcing myself to take the day off before my long run each week. I started stretching more in the morning. And I slowed down my long training runs by just ten seconds per mile.

At 60 years old you just have to break 4 hours to qualify for the Boston Marathon.[54] In 2009 I ran the Boston Marathon in 3 hours, 36 minutes and 58 seconds. I realized I could easily slow down 10-15 seconds a mile and still qualify with plenty of time to

---

[54]This qualifying time was reduced to 3 hours and 55 minutes in 2012

spare for the following year.

I made these changes in December, 2009 and by February 2010, I found that my quads, knees, Achilles and back were not hurting as much as they had been. I don't know how much longer I will be able to continue running marathons, but I have learned that each year brings changes to what I face and with each new set of conditions I always have options that I may not have tried before—and I always need to be ready to change my rules again. There is no single best way to train for a marathon, just like there is no single best way to get better at anything you want to get better at. The process of self-discovery never ends.

I finished the Vermont City Marathon in 2008 in 3 hours, 37 minutes, and 53 seconds and was very surprised by my time. I had thought I was running much slower for those last 14 miles where I continued to push through a cramp in my right calf. By the time I finished the cramp was gone and I had no lingering after-effects. I ran five miles the next day. When I asked my wife how I looked when she saw me running those last 14 miles while I was fighting my calf cramp most of the way she said I looked fine and she had no idea what was going on inside me. I finished the 2010 Boston Marathon in 3 hours, 37 minutes, and 02 seconds. I finished the 2011 Boston Marathon in 3 hours, 37 minutes and 58 seconds.

# Appendix A – Building a Library of Patterns

Consider keeping a history of common patterns in your organization, including the results of analysis of each pattern, actions taken and results. This data can then be used to help build performance models as discussed in Chapter Ten. Then periodically go back and review this data looking at actions taken and past results. Refreshing this information before an important meeting can help you prepare to make more effective decisions. Refer to table below.

| Common Patterns | Root Cause Analysis | Actions taken and results |
|---|---|---|
| 1 Dissatisfied customer project X | ….. | …… |
| 2 Excessive acceptance test defects project Y | …. | … |
| 3 Ambiguous or incomplete requirements | | |
| 4 Stakekholders too busy to collaborate | | |

Following are some questions to ask that can help you establish your own patterns to aid your performance.

- *Do you know the specific patterns that affect performance that commonly occur in your organization, and the common signals that indicate one of those patterns is starting?*
- *Have you recorded them so you can go back and see how often these patterns are occurring?*

- *Do you know what actions to take when you sense those signals because those actions have worked in the past under similar conditions to help sustain high performance?*
- *Do you know what actions to avoid when you sense those signals because they have failed in the past under similar conditions leading to poor or reduced performance?*
- *Are you recording how often these actions that should be taken are actually taken and followed up?*
- *Do you know how to keep the signals and related positive and negative patterns fresh in the front of your mind just when the signals are most likely to occur?*

When your data shows clear patterns that call for specific actions, you can add hints to your practices to remind yourself to take the right action at the right time.

---

# Appendix B – Performance Models Aiding Practitioner Daily Decisions

Historically, many organizations that have applied the CMMI high maturity practices have not used their data to aid practitioners in every day decisions. But there is no reason such an approach could not be employed, and there are examples of CMMI level 5 organizations that have succeeded in improving their performance using such an approach.

If your organization uses the CMMI model, you can use your pattern data together with Causal Analysis and Resolution to raise the visibility of positive and negative patterns in your organization.

## Using Causal Analysis and Resolution (CAR) and Performance Models To Raise the Visibility of Positive and Negative Patterns

The purpose of Causal Analysis and Resolution (CAR), a CMMI Level 5 process area, is to identify causes of selected outcomes, both positive and negative, and take action to improve process performance. Keeping historical data on common patterns and creating related performance models can be an effective approach to help determine which positive patterns are working best to sustain or improve performance in your organization, and which negative patterns may be hurting your performance the most.[55]

Referring to the hypothetical data in Figure App B-1 one could conclude that pattern X has a definite correlation to improved customer satisfaction. An organization could use this data to help in making a decision related to investing more in pattern X to improve its organizational performance related to customer satisfaction. On the other hand the organization could conclude that it isn't worth investing in pattern Y if the

---

[55] Maintaining performance models at the personal level is a best practice advocated by the Personal Software Process (PSP) developed by Watts Humphrey

objective of pattern Y is to improve cost and schedule management because the data indicates no correlation.

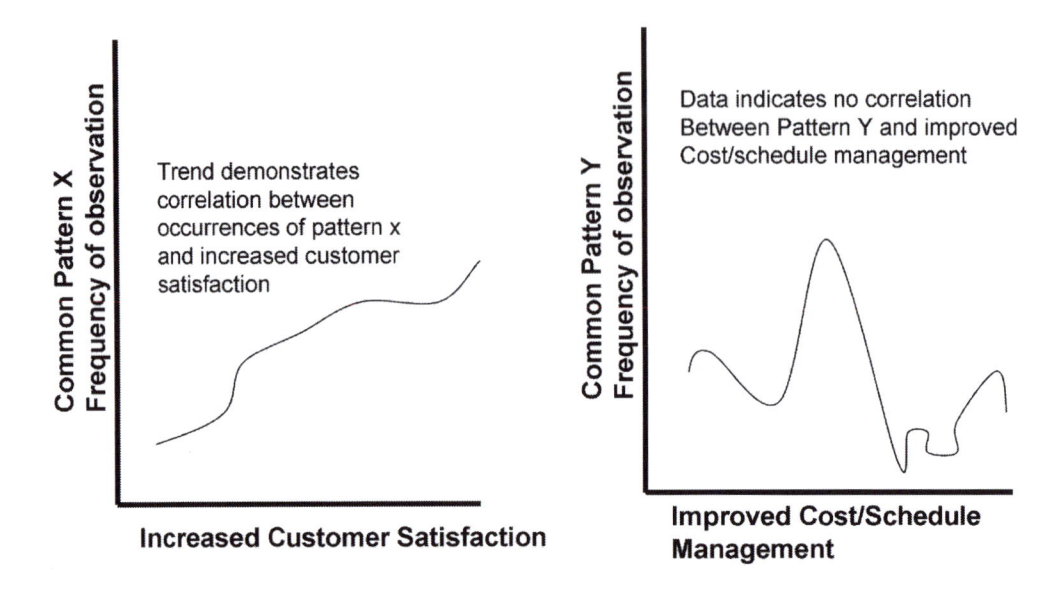

Figure App B-1 Performance Models Demonstrating Positive & Negative Patterns

# Appendix C – Simple Example of Second Level Checkpoint

The usefulness of second level checkpoints using objective data in business can be seen through one of my recent experiences helping a client implement a new set of processes. My client told me he thought the processes we developed were the right processes, but he knew he needed a more objective assessment. After some discussion the objective measures agreed to were *"customer feedback"* and the *"number of discrepancies found during final customer test."*

What is interesting about these measures is that they are *objective* in that they both come from an external perspective–the customer's view point, [56] and they tie to the real goal. The insight we learned from this experience was that the ultimate goal wasn't to see if the organization was just following their new processes, but rather to ensure we had developed the right processes to help the organization achieve their real goal which was a satisfied customer. Continually listening to your customer feedback and monitoring discrepancies found during final customer test should lead to more questions, analysis and actions. Refer to table below.

| Objective data | Question, Action |
| --- | --- |
| Dissatisfied Customer A | Pattern seen before? Action XYZ worked in similar situation |

Second level checkpoints help identify defects early related to *"how well"* we are performing. These types of defects (e.g. dissatisfied customer, too many defects during final acceptance testing, failing to raise a risk at the right time) often escape when using first level checks alone, and they are often the most critical type when it comes to high value performance improvements.

---

[56]Objective measures can be achieved through an external source, or by looking at existing data from a different perspective.

# Appendix D – Comparing Performance Improvement Frameworks

Lean Six Sigma, CMMI, Scrum and Essence can each be viewed as performance improvement frameworks. By understanding the strengths and weaknesses of each you can learn how each of these aids can be used most effectively to help you with the challenges you face.

*Lean Six Sigma*

First, Lean Six Sigma is a toolkit with many techniques and you aren't expected to use them all on any given improvement effort, but rather you are expected to pick the ones relevant to your situation. Lean Six Sigma is best at attacking a specific problem providing techniques to help locate root causes and put resolutions in place that can then be monitored to ensure performance goals are met.

Its strength lies in its many proven *"how-to"* techniques. Its weakness lies in the time it takes to learn and become an expert in its many techniques. It takes approximately 200 hours of intense study, followed by a five hour timed examination to achieve a Lean Six Sigma Black Belt, and the black belt competency level is often required to apply the Lean Six Sigma techniques correctly.

*CMMI*

The CMMI is a comprehensive process improvement reference model which is strong at helping organizations assess the current state of their processes laying the foundation for an improvement plan. It is a general model that focuses on what you should be doing, and where you may have gaps, but it does not dictate, or provide help in, how to go about filling your gaps or improving, or in making your processes more lean and cost competitive.

This is both a strength and weakness of the model. It is a strength in that it makes the model widely useable as an assessment and planning tool. But it is a weakness in that it does not provide the "how to" techniques, such as those provided by the Lean Six Sigma toolkit to locate root causes and monitor improvements. Another weakness of the CMMI is the time required to understand the model and apply it. The CMMI-Development guidelines book is over six hundred pages, and requires process improvement professional expertise to apply.

*Scrum*

Scrum can be viewed as a Project Management Framework. Its strength is its simplicity and wide appeal. It is easy to learn enough to get started quickly, but very difficult to master because it lacks "how to" specifics for many essential practices. Other weaknesses include the fact that it contains no technical practices and it doesn't address organizational improvement issues such as sharing across projects.

*Essence*

The Essence approach can be viewed as a Software Engineering Framework that is different in multiple ways from the previous three frameworks discussed. First, Essence provides a small set of essentials, and it intentionally contains no best practices or techniques. Its focus, like Scrum, is at the endeavor (e.g. project) level.

The strengths of the Essence framework include a capability to add as many or as few of your own practices as you see fit. You can also use the framework to assess your situation, but the assessment is relative to a specific endeavor, not an organization as is the focus of the CMMI framework.

Another strength of the Essence framework is that it is for the professional practitioner, not the process engineer and it is expected to be used during an endeavor, not just in preparation for endeavors. A strength of the Essence approach is that the basic model can be learned quickly and professionals can begin to apply it almost immediately to help in their work effort. But it also requires practice to learn to apply it effectively and consistently on your specific project. For more information refer to Part III in this book.

# Appendix E – More Evidence Supporting Need for Integrated Practice

Daniel Kahneman[57] in his widely acclaimed book, "Thinking Fast and Slow" [19] provides numerous examples where humans tend to jump to conclusions based on inadequate information. Kahneman uses the analogy of humans operating like two systems. System 1 is the fast thinker within us who doesn't like uncertainty and is quick to jump to conclusions with minimal supporting evidence. System 2 is the slow thinker within us that is more logical, but unfortunately, as Kahneman lets us know, the System 2 inside most of us is also lazy too often giving in to System 1's rash desires.

Kahneman tells us, the System 1 within us is constantly pushing us to ignore anything that gives us a sense of conflict or uncertainty. Because of the way System 1 operates Kahneman points out that most humans are not naturally good intuitive statistical thinkers.

When I read Kahneman's book it reminded me of Jonah Lehrer's book (discussed in Chapter Eight) "How We Decide" [20]. This was more compelling evidence for the need to find a way to integrate both our fast and slow thinking sides (or in Lehrer's terms our emotional and rational sides) to help us make both better informed and rapid decisions.

In CMMI terms, the better informed side is what process performance baselines and models are about. But such data will not be useful to us when it comes to performance if it isn't integrated with our needs on the job when it comes to decisions.

In Part III of Kahneman's book (Chapter 22) he explains what we can do to help our quick thinking be more accurate. His answer has distinct similarities to what we now know about how our great athletes learn to excel. You can improve your quick, or

---
[57] Dan Kahneman, a cognitive psychologist and professor at Princeton won the Nobel Prize for Economics in 2002

intuitive, thinking through practice. Kahneman tells us that experts improve through practice, or mental simulation, learning to recognize common situations rapidly. He also tells us that intuition is nothing more and nothing less than recognition.

You first must learn to recognize common situations, and then learn the appropriate response. This takes practice. As you find patterns consider collecting hints, and examples to help you keep your practices useful and current. Then continually practice your practices with your latest and best data. You can be fast and agile, but you want to make sure you are accurate too.

# Appendix F – Cross-Reference CMMI Specific Topics in Book

| CMMI Topics | Location |
|---|---|
| What this Book is About | Introduction |
| Common Measurement Mistakes | Chapter Three |
| Collect, Analyze, Act: Fundamental... | Chapter Four |
| What's Happening Today in CMMI Level 5.. | Chapter Four |
| First Level Checkpoints... | Chapter Five |
| Challenges Related to Effective Root Cause.. | Chapter Six |
| Spreading Positive Performance... | Chapter Nine |
| The Right Patterns... | Chapter Ten |
| Implementing CMMI High Maturity... | Chapter Ten |
| Increasing Practitioner Involvement... | Chapter Ten |
| Sidebar: Motivation for increasing... | Chapter Ten |
| Statistical Process Control simplified... | Chapter Eleven |
| Examples Statistical Process Control | Chapter Eleven |
| Comparing Company A and B's Approach | Chapter Eleven |
| Why an Unstable Process Could Help... | Chapter Eleven |
| Spreading Positive Performance... | Chapter Twelve |
| Sidebar: CMMI tip on objectives | Chapter Twelve |
| Using Performance Models... | Appendix B |
| Comparing Performance... | Appendix D |

# Appendix G – Cross-Reference Lean Six Sigma Specific Topics in Book

| Lean Six Sigma Topics | Location |
|---|---|
| Sidebar:Lean techniques, bottlenecks… | Chapter Two |
| Sidebar: How Lean Six Sigma can help | Chapter Three |
| Collect, Analyze, Act: Fundamental… | Chapter Four |
| Wrong Way to Conduct Root Cause Analysis | Chapter Six |
| Sidebar: Touch-point to Lean Process… | Chapter Eight |
| Sidebar: How Lean Six Sigma can help | Chapter Ten |
| Statistical Process Control Simplified | Chapter Eleven |
| Ask Why Five Times to Guide Search… | Chapter Twelve |
| Comparing Performance Improvement Frameworks | Appendix D |

# Appendix H – Cross-Reference Agile Specific Topics in Book

| Agile Topics | Location |
|---|---|
| Sidebar:Agile Retrospectives… | Chapter Four |
| Collect, Analyze, Act: Fundamental… | Chapter Four |
| Sidebar: Scenarios helping agile teams | Chapter Nine |
| Sidebar: Structured Real Story helping… | Chapter Ten |
| Increasing Practitioner's Involvment | Chapter Ten |
| Practical High Maturity/Agile Retrospectives | Chapter Eleven |
| Second Example: Empowering Your scrum Team | Chapter Eleven |
| Pattern 2 for Better Decisions… | Chapter Fourteen |
| Comparing Performance Improvement | Appendix D |
| Pattern 7: Using Data to aid Improvement | Appendix P |
| Pattern 9: Assisting Self-Direction | Appendix P |
| Sidebar: Not just for managers… | Appendix Q |

# Appendix I – Cross-Reference Practitioner Scenarios

| Practitioner Challenge/Scenarios | Location |
|---|---|
| Procurement Case | Ch 2,6,7,12 |
| Common Measurement Mistakes | Ch 3,10 |
| Business Case Sustained Performance | Ch 4 |
| Small change requiring practice | Ch 4 |
| First Level in Business… | Ch 5 |
| Examples improving quality checks.. | Ch 5 |
| Planning challenge… | Ch 6 |
| Business example 1st level check… | Ch 6 |
| Business example common interactions… | Ch 6 |
| Big picture feel planning… | Ch 6 |
| Second level check… | Ch 7 |
| Wrong Can-do vision | Ch 8 |
| Software developer pattern examples | Ch 9 |
| Listen for technique | Ch 9 |
| Multiple pattern examples… | Ch 9 |
| Agile team scenarios… | Ch 9 |
| Dictated schedule challenge | Ch 10 |
| Frustrated Process engineer… | Ch 11 |
| Empowering your Scrum team… | Ch 11 |
| Multiple examples practitioner pain | Ch 13 |
| Overly simplified process guidance | Ch 13 |
| Multiple examples of Essence checklists | Ch 13 |
| Multiple pattern examples | Ch 14 |
| Multiple pattern examples | App P |
| Pattern example | App Q |

# Appendix J – Intentionally left blank

# Appendix K – About the Term Pattern in this Book and An Analogy

In this book I use the term pattern in two different, but consistent ways.

A pattern, in the general sense, as I use the term throughout this book, is an abstraction of a common situation that occurs in a specific context (could be good or bad) with one or more possible alternative solutions.

A pattern in the Essence language is an arrangement of language elements (e.g. Alphas, Alpha states, competencies…) into meaningful structures. The patterns I define in Part III of this book, and in Appendix P and Q, using the Essence language pattern construct are intended to aid practitioner's by helping them recall common situations that occur in their specific work environment along with possible alternative decisions they could make when faced with similar situations.

## An Analogy to aid Understanding of Patterns

A Use Case slice can be thought of as a single instance through a Use Case. Similarly, a common scenario can be thought of as a single instance that occurs when using a practice. I have defined pattern in this book to mean an abstraction of a common scenario. To aid understanding, the following analogy could be useful.

Practices are to the organization what Use Cases are to a Software System. Common scenarios are to Practices what Use Case Slices are to Use Cases. Patterns are abstractions of common scenarios that can be used to help practitioners recognize common situations and make better decisions. Refer to Figure App K-1.

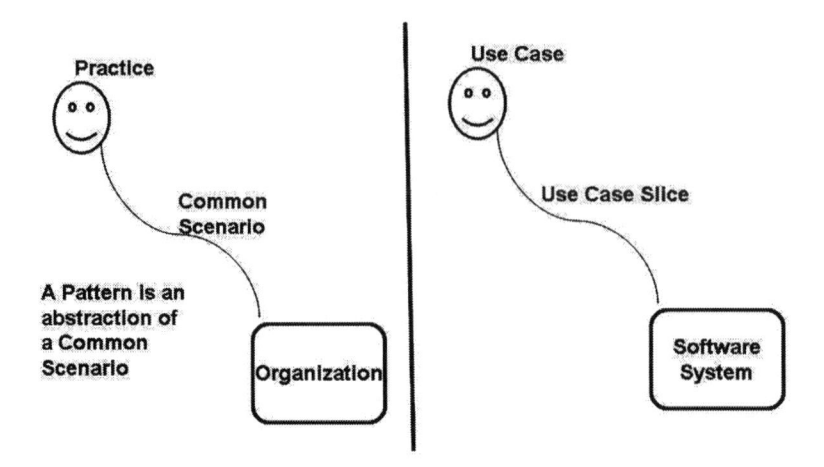

# Appendix K – About the Term Pattern in this Book and An Analogy

Figure App K-1 An Analogy to Aid Understanding of Patterns

We develop and test software systems incrementally with use case slices to ensure the software system meets its requirements. Similarly we improve the competencies of our people in our organization by sharing with our less experienced personnel common patterns that our experts know.

# Appendix L – Why Risk Is Not an Essence Thinking Framework Alpha

Wikipedia tells us that risk is the potential that a chosen action or activity (including the choice of inaction) will lead to a loss (an undesirable outcome).[58] While the importance of risk management is widely accepted, many organizations have struggled to implement effective risk management practices. This leads us to ask:

*How can the Essence thinking framework help software professionals implement effective risk management strategies?*

Since effective risk management is essential to all software endeavors one might conclude risk should be included inside the Essence framework. But then we must ask:

*Where exactly does it belong inside the framework? Should it be part of a specific Alpha?*

If so, *what Alpha should it be related to?*

The Alphas represent the most important things that need to be monitored, and progressed, and while they are not independent of each other, they can each be progressed based on their own distinct states and checklists. Risk is different. The risk of a software endeavor is directly affected by the health of all of the Alphas.

When a group of software professionals in a large US defense company was asked if they could see the value of the Essence framework after the framework idea had been explained to them, one experienced software professional commented on how they saw the complete framework as a potential aid to implement an effective risk management practice.

---

[58]The use of agile practices has led to the identification of two types of risk: Negative (as in the Wikipedia reference) and positive. Positive risks relate to unexpected, serendipitous events that are good for the endeavor.

As a quick example, most software endeavors focus heavily on the Requirements and Software System Alphas to determine progress. These Alphas can be progressed even when risk exists that the ultimate endeavor goal may not be achieved. In these cases, keeping an eye on the health of the Opportunity, Work, Team and Way of Working Alphas can often help to surface potential problems (e.g. risks) early enough to take practical actions to help ensure the software endeavor's ultimate success.

Keep in mind, the framework contains no practices. However, since risk can surface in any of the Alpha elements, the framework along with its states and checklists can be viewed as an aid to remind people where to look for risk, and as an aid to help find effective risk mitigation approaches.

By keeping risk out of the framework each endeavor is free to use the framework to establish their own explicit risk management practice, or to use the framework to aid risk management in a tacit manner. This is consistent with the idea that the framework is practice agnostic, supporting all approaches to effectively manage software development and its associated risks.

# Appendix M – Questions About the Essence Framework

*Why is the Team Alpha separate from the Stakeholders Alpha? Aren't team members stakeholders too?*

This is a common question when people first start examining the Essence thinking framework. The answer is, yes, team members are stakeholders too. Using the Alphas requires a paradigm shift from the way many practitioners think today. The Alphas should not be viewed as a physical model of your project. Rather they are critical indicators representing the things that are most important to monitor and progress.

*Doesn't the sequential nature of the Alpha states require a waterfall approach?*

This is a common misunderstanding. It is true that the Alphas do progress from state to state through a sequence of actions taken by development teams in response to the current state. But it is wrong to view this as a strictly sequential assembly-line type process.

The Alphas provide a set of critical indicators that help the team take action including when their endeavor starts to get out of kilter. The Alphas help the team tune their behaviors and their communications to keep the endeavor running smoothly and predictably. As conditions on a project change you can fall back into previous states, and depending on your chosen life cycle and practices you can iterate many times through a sequence of states. Furthermore, different parts of your system can be in different states at the same time with the same Alpha.

*Why was the term Alpha chosen, rather than using a more traditional term?*

We chose the term to highlight the fact that the Alphas are something new and we wanted to differentiate them from traditional approaches to measure progress and health by measuring physical documents or artifacts. The Alphas should not be viewed as simply abstract artifacts. Rather, the Alphas are the critical indicators that help a team assess the progress and health of their endeavor. They always exist on all software endeavors regardless of the practices being employed and degree of documentation.

*Why is hardware included in the definition of the Software System Alpha?*

The Essence framework uses the term software system rather than software because software engineering results in more than just a piece of software. Whilst the value may well come from the software, a working software system depends on the combination of software, hardware and data to fulfill the requirements.

# Appendix N – A Caution When Using the Essence Framework

There is some similarity between the Essence Alpha checklists and specific and generic practices in the CMMI model. In the CMMI model specific and generic practices are expected, but they are no assurance of achieving the CMMI Goals and it is the goals which are most important to performance. In the past, it has not been uncommon for organizations to become overly focused on theses specific and generic practices failing to consider their own organizational context thereby missing the goal. The same risk exists as organizations begin applying the Essence approach.

While each Essence Alpha state checklist represents something that can help you achieve your performance objective, all of the checklist items taken together will not ensure you achieve your real objective. Furthermore, depending on your specific situation some checklist items may be critical to achieving your objective, while others may take on far less significance.

As an example, refer to *Pattern 3 for Better Decisions: Keeping An Opportunity Alive and Healthy*, in Chapter 14 of this book. Alphas, their states and checklists are based on past experience, but they will never be perfect because future situations never reflect exactly past situations.

The value of the Essence checklist items is that they have been validated by practitioners across a broad range of software applications, University professors from around the world, and industry leaders all as part of volunteers on the SEMAT effort. Nevertheless, while someone needs to be keeping an eye on, and thinking about, each of the Alphas, and their checklists, the effort required for each will vary dependent on the specific circumstances of your endeavor.

# Appendix O – Intentionally left blank

# Appendix P – 6 Additional Essence Patterns

As discussed in Appendix K, a pattern is an abstraction of a common scenario. In Chapter 14 we discussed 3 patterns. In this appendix we provide 6 additional patterns based on common scenarios described that provide more extensive examples of the use of the Essence framework to help practitioners make better decisions.

## Pattern 4: Reasoning Through a Requirements Dilemma

*Some believe software developers just need good tools, a workstation and to be left alone to write their code. While this may in fact be true in many cases there is a scenario I run into when talking to software developers that occurs far more often than many of us would like to believe– and it relates directly to a software development team's performance. What the team needs to decide in this scenario is whether or not this situation is really a problem and if so what action they should take. This scenario can help you understand how the Essence Requirements Alpha and checklists can potentially help your team's performance by helping them make better on the job decisions. This example should be of particular interest to developers and students.*

I work with a number of organizations that develop software for the U.S. Department of Defense (DoD). Some of these companies develop simulation software to help train our service personnel. One of the developers I worked with was frustrated trying to understand one of his customer's requirements. He explained the problem by stating that when he asked the customer for more details about what he needed, the customer just said:

*"We need simulated missiles that fly high and go fast."*

Developers know they can't implement "fly high, and go fast." But what happens when they ask how high and how fast, and the customer responds:

*"I don't know. Just build something and then we'll tell you if its good enough."*

Often, in today's world, this is the scenario that plays out, and so the developer build something, shows it to the customer and the customer says:

*"That's not what I had in mind. Try again."*

So, is this a problem or not?

Inadequate requirements may be the most common and costly problem software developers face, but what options do they have when faced with this common scenario? Software professionals can quickly refresh what they need to know about requirements essentials through the Requirements Alpha and its states. Refer to Figure P4-1.

**Requirements**

- Conceived
- Bounded
- Coherent
- Acceptable
- Addressed
- Fulfilled

What the software system must do to address the opportunity and satisfy the stakeholders.

- Good Requirements meets real needs
- Good Requirements has clear scope
- Good Requirements are coherent and well organized
- Good Requirements help drive development

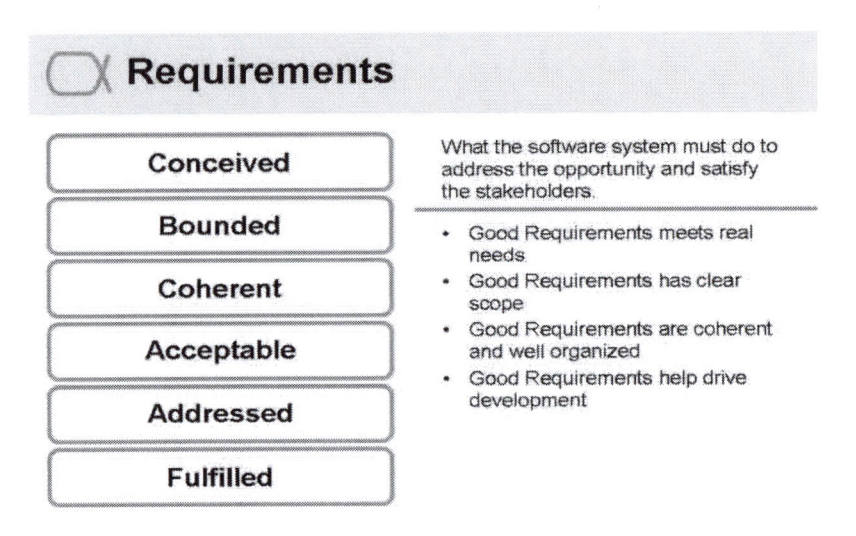

**Figure P4-1 The Requirements Alpha**

The Requirements Alpha is defined as:

- *What the software system must do to address the opportunity and satisfy the stakeholders.*

What every software professional should understand is the intent of the states of this Alpha and how they can help anyone facing similar challenges reason through their options and consequences to arrive at the best decision.

As an example, let's look at two specific *states* of the Requirements Alpha and discuss how a software developer might use this information to help decide how to handle this common scenario.

The *Acceptable* state is defined as:

- *The requirements describe a system that is acceptable to the stakeholders.*

The *Addressed* state is defined as:

- *Enough of the requirements have been addressed to satisfy the need for a new system in a way that is acceptable to the stakeholders.*

Refer to Figure P4-2.

**Requirements**

**Acceptable**
- Requirements adequately describe an acceptable solution to stakeholders
- Rate of change to agreed requirements is low and under control

4 / 6

**Requirements**

**Addressed**
- System implementing requirements is worth making operational
- Enough requirements are implemented

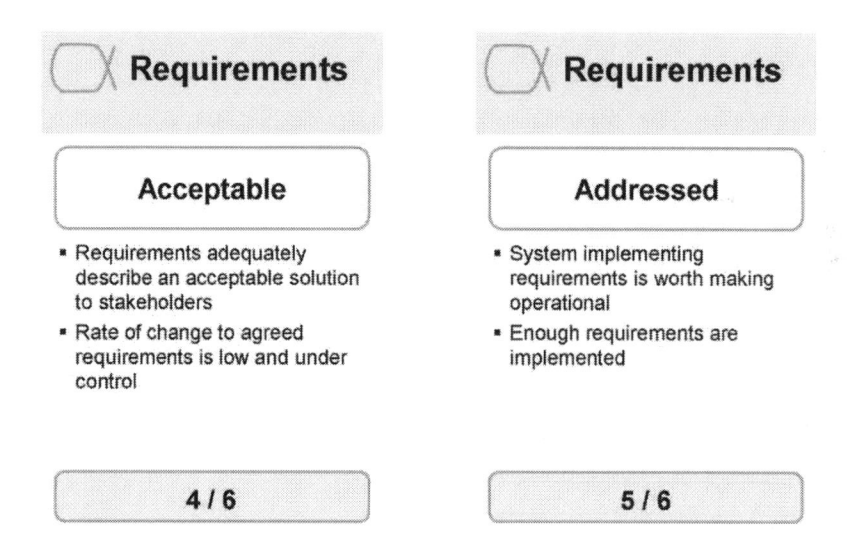

5 / 6

Figure P4-2 Requirements Alpha Acceptable, and Addressed States

Stakeholder and development team collaboration is essential to ensure we are converging on an agreed to set of requirements that are *Acceptable*. But if we get ahead of ourselves working hard on getting to the *Addressed* state before we have achieved the *Acceptable* state there is significant risk of rework as we saw in our example.

On the other hand, addressing some requirements early so we can demonstrate parts of the system to the user can help us to move forward toward an *Acceptable* set of requirements by helping the customer discover their own requirements. However, the team should understand that there are tradeoffs to consider when making a decision including potential risks of proceeding too far without an acceptable set of requirements.

Some people when they first look at Essence think the framework requires a sequential waterfall approach. In other words, they think they need to get ALL the requirements to the Acceptable state, before they work on getting any of them to the Addressed state. This is an incorrect way to view the Alpha states.

For example, different parts of your requirements can be in different Alpha states at the same time, and you can cycle back through the same state multiple times. You can also regress. That is, you could achieve a state, but at a later time in your endeavor discover you are no longer meeting all the checklists. How a team works through the Alpha states depends on a team's chosen life cycle and practices. The Essence approach is agnostic to your team's chosen method.

Some of the questions the Requirements Alpha checklist items will lead a practitioner to ask include:

- *Is it clear what success is for the new system?*
- *Do the stakeholders have a shared understanding of the extent of the new system?*
- *Do we know the essential characteristics of the new system?*
- *Does the team understand what has to be delivered?*
- *Do we have a mechanism in place to manage changes to the requirements?*

By asking these questions it can help team members decide whether the state of their requirements (or the state of a specific part of their requirements) is a problem or not that requires them to take some immediate action. The answer often depends on how your team and the customer are working together.

For example, if you are using an agile approach, and your customer is collaborative and understands the impact of changing requirements, then you may be fine.

However, if you have a fixed cost and schedule and a non-collaborative customer, then not getting to the Acceptable state could present a high risk. This is an example of using the Essence framework as a thinking framework. It helps you keep your options and consequences fresh in your mind when you need it most, and it helps you ask yourself the right questions leading to the best decision given your current situation.

Note in this case we actually have more than one goal state occurring at the same time. This is not unusual, especially on projects using iterative and agile approaches. Keep in mind the framework is practice and method agnostic. It can help practitioners assess their situation and make sound decisions regardless of their chosen life cycle, practices and method. Refer to Figure P4-3.

**State Achieved**

Requirements
Coherent
- Described requirements provide coherent picture of the system
- Conflicting requirements separated
- Important usage scenarios explained
- Priority of requirements clear

3 / 6

**Next Goal State (s)**

Requirements
Acceptable
- Requirements adequately describe an acceptable solution to stakeholders
- Rate of change to agreed requirements is low and under control

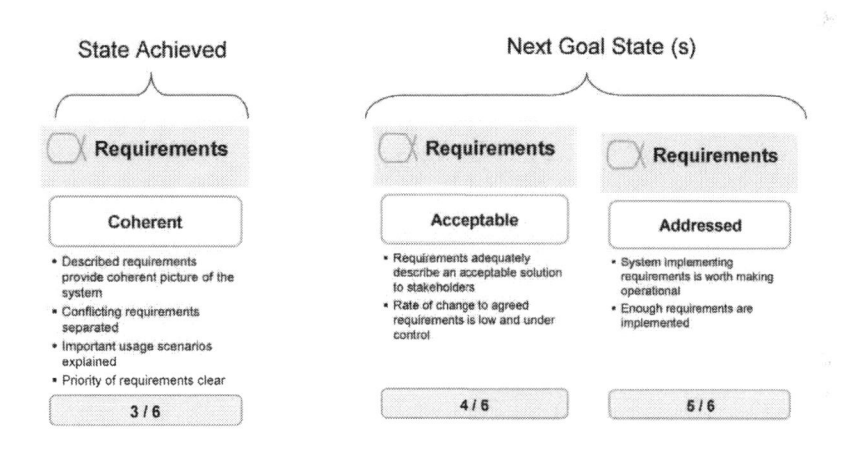

4 / 6

Requirements
Addressed
- System implementing requirements is worth making operational
- Enough requirements are implemented

5 / 6

Figure P4-3 Requirements Alpha Target States

To ask critical questions and reach sound solutions with respect to the state of your requirements requires an individual with the *Analysis* competency. This competency is defined as:

- *A competency that encapsulates the ability to understand opportunities and their related stakeholder needs, and transform them into an agreed and consistent set of requirements.*

Refer to Figure P4-4 for a diagram of the Reasoning Through a Requirements Dilemma Pattern. This pattern can remind developers of their options when facing difficulty getting stakeholder agreement on acceptable requirements.

**Pattern Trigger:**
Difficulty getting stakeholder agreement on acceptable requirements.

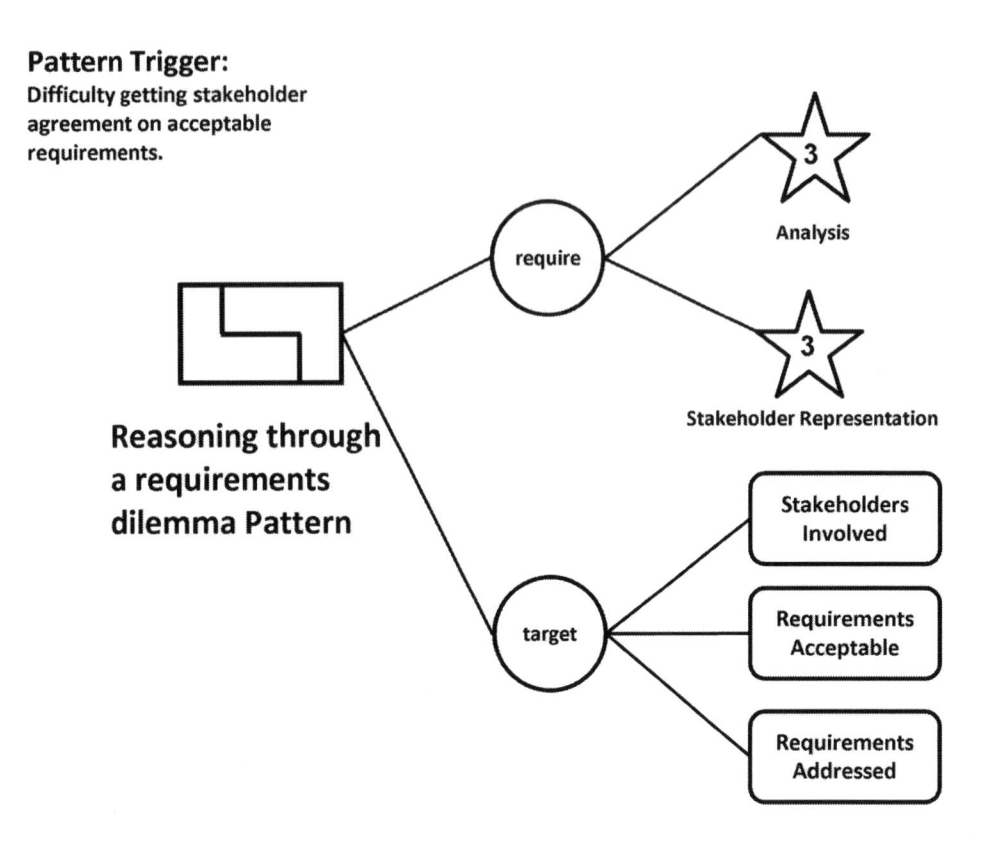

Figure P4-4 Reasoning Through a Requirements Dilemma Pattern

# Pattern 5: Reasoning Through a Lack of Data Challenge

*This example is similar to our previous example in that something is impacting the team's ability to get their work done, but in this case it may be more serious. Because the consequences are not evident to the new project manager in time, you will see how it costs his company considerable profit. You will learn in this example how the Essence thinking framework could have helped our project manager make a different and potentially better decision that could have saved his company from their lost profit. This example should be of particular interest to project managers.*

On the LEM project Don's company was subcontracting component testing. Critical test data was required to be supplied through a customer agency after contract award. Don was a new project manager and he wasn't sure what he should do when his point of contact at the customer agency kept putting off his repeated requests for the contractually required data. Don knew, as the prime contractor, it was his responsibility to ensure all testing was completed satisfactorily before final acceptance, but he didn't want to upset his customer especially on his first assignment as a project manager.

Unfortunately, his subcontractor wasn't getting the testing done and they were falling further and further behind schedule, as the subcontractor kept blaming the lack of test data for the slipping schedule. Eventually the data arrived and Don made the decision to proceed by approving additional budget for his subcontractor to complete the testing. But since this project was at a fixed cost to the prime, it resulted in significant lost profit for his company.

*Was there a better decision that Don could have made, and if so when should he have made it to save his company's lost profit and still keep his customer satisfied and happy?*

Lets now look at a few of the relevant Alphas that might have helped Don make a different and potentially better decision. Lets start with the Software System Alpha.

One of the states of the Software System Alpha is *Usable* which is defined as:

- *the system is usable and demonstrates all of the quality characteristics required of an operational system.*

Refer to Figure P5-1.

## Software System

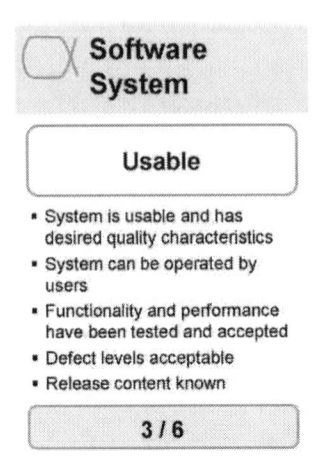

**Usable**

- System is usable and has desired quality characteristics
- System can be operated by users
- Functionality and performance have been tested and accepted
- Defect levels acceptable
- Release content known

**3 / 6**

**Figure P5-1 Software System Alpha Usable State**

Don knew he wasn't getting to the Usable state as fast as he needed to. Specifically, this Alpha would have alerted him that he was having trouble with the checklist item that states:

- *The functionality of the system has been tested.*

But just knowing this wouldn't have been enough information to alert him to what other options he might have had. Often the issues faced, especially on complex endeavors, cross multiple Alphas. By knowing which Alphas to keep a close eye on in certain situations practitioners can be better prepared to respond appropriately to a common situation such as the one Don was facing in our story. Specifically, in this case, the Alpha of particular interest is the Stakeholders Alpha. Refer to Figure P5-2.

## Stakeholders

**Recognized**
**Represented**
**Involved**
**In Agreement**
**Satisfied for Deployment**
**Satisfied in Use**

The people, groups, or organizations who affect or are affected by a software system.

- Healthy stakeholders represent groups or organizations affected by the software system
- Healthy stakeholder representatives carry out their agreed to responsibilities
- Healthy stakeholder representatives cooperate to reach agreement
- Healthy stakeholders are satisfied with the use of the software system

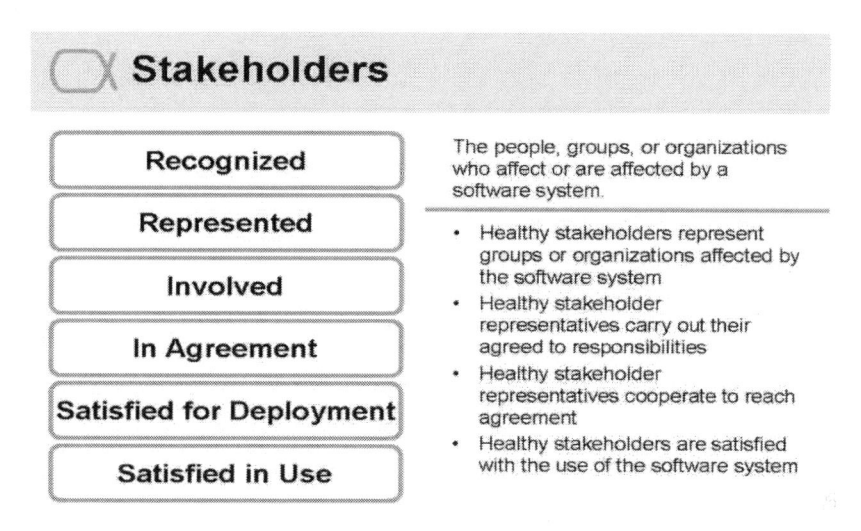

Figure P5-2 Stakeholders Alpha

The Stakeholders Alpha is defined as:

- *The people, groups, or organizations that affect, or are affected, by a software system.*

Lets now look closer at two states of the Stakeholders Alpha, *Represented*, and *Involved*.

Refer to Figure P5-3 and Figure P5-4.

**Stakeholders**

**Represented**
- Stakeholder representatives appointed
- Stakeholder representatives agreed to take on responsibilities & authorized
- Collaboration approach agreed
- Representatives respect team way of working

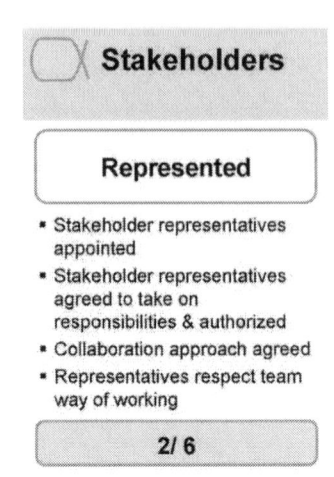

2/ 6

Figure P5-3 Stakeholder Alpha Represented State

The Represented state is defined as:

- *the mechanisms for involving stakeholders are agreed and the stakeholder representatives have been appointed.*

**Stakeholders**

**Involved**
- Stakeholder representatives carry out responsibilities
- Stakeholder representatives provide feedback & take part in decisions in timely way
- Stakeholder representatives promptly communicate to stakeholder group

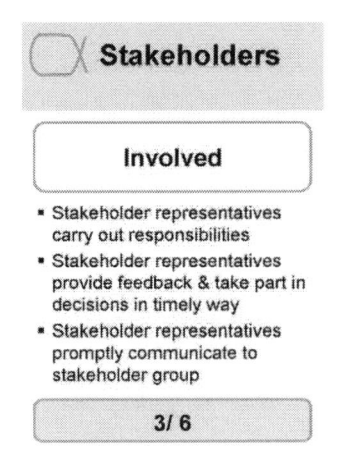

3/ 6

Figure P5-4 Stakeholder Alpha Involved State

The Involved state is defined as:

- *the stakeholder representatives are actively involved in the work and fulfilling their responsibilities.*

A related checklist item in the Represented state is:

- *the stakeholders have agreed to take on their responsibilities*

And a related checklist item in the Involved state is:

- *the stakeholders are assisting the team in accordance with their responsibilities.*

A checklist item from the Software System Alpha Usable state alerted Don that he had a problem, but it didn't help him understand what options he had to solve it. If Don had been keeping an eye on the progress and health of his Stakeholders Alpha it might have led him to ask the following questions:

- *Have all the key stakeholders been appointed?*
- *Have they accepted their responsibilities?*
- *Are they actively involved in the work fulfilling their responsibilities?*

Refer to Figure P5-5 to see the state of the Stakeholders Alpha that had been achieved on the LEM project, and the next goal states.

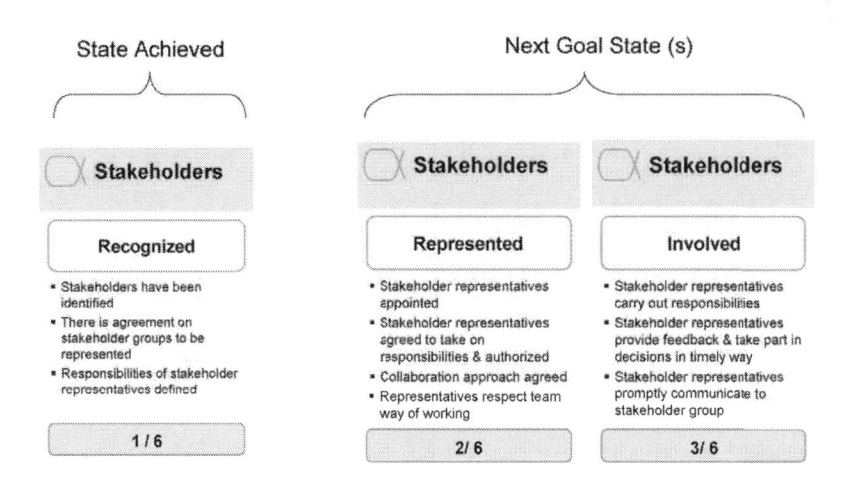

Figure P5-5 Progress and Health Assessment of Stakeholder Alpha

Recall that the definition of the Stakeholders Alpha includes both people who are affected, and affect, the software system. Stakeholders are not just customers. They are anyone who can affect the software system as well. This includes personnel who must supply data. Asking the questions identified above could have led Don to communicate his schedule problem to his customer. The customer might have been able to help Don get the needed test data sooner.

If the customer couldn't help, at least he would have been warned of the problem, and then once Don realized he had a budget shortfall he could have gone back to the customer for the additional budget, rather than proceed and cost his company a good part of their profit.

What Don needed was a little more experience to help with his *management and leadership* competencies.

The *Management* competency is defined as:

- *this competency encapsulates the ability to coordinate, plan and track the work done by a team.*

It is not uncommon, especially on complex efforts, when the work isn't getting done by the team that the root cause comes back to a combination of management and leadership.

Refer to Figure P5-6 for a diagram of the *reasoning through a lack of data* pattern that could be used to remind an inexperienced project manager of the need for represented and involved stakeholders when they have external dependencies that are impacting their team's performance.

**Pattern Trigger:**
External dependencies impacting team progress.

**Reasoning through a lack of data pattern**

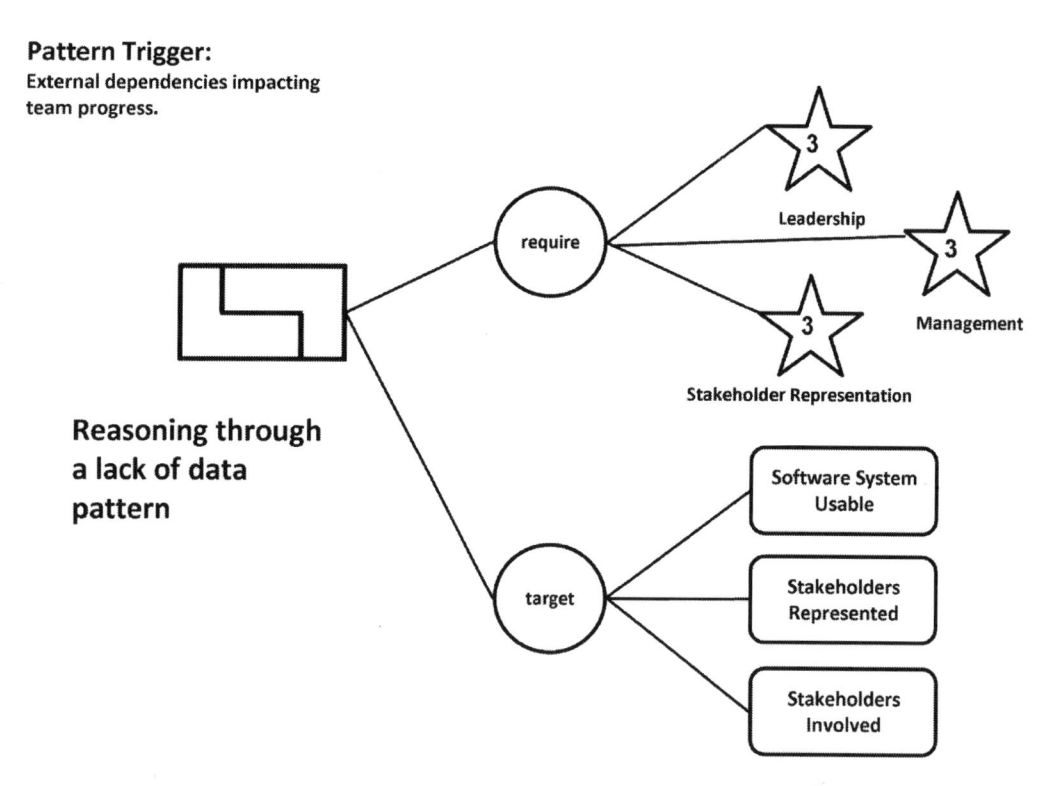

Figure P5-6 Reasoning Through a Lack of Data Pattern

In hindsight Don recognized this would have been a better decision, but under the stress of his project, and especially being a new project manager, he missed the signal. If he had a little more management experience it is possible he wouldn't have missed this signal. This is where the Essence approach can help.

If Don had been using the Essence thinking framework, it could have reminded him that he had other options. The framework won't give you specific answers to your challenges, but your own patterns built on top of the framework can help by reminding you, or your people, of similar situations along with related options and potential consequences.

> **Sidebar: Competency**
>
> In many situations when poor decisions are made, the underlying issue often comes down to the competency of your staff. Project leaders need to ask: *"Do we have the right competencies, given the complexity of this endeavor?"* The Essence framework provides necessary essentials, but it is not sufficient to address the needs of many challenging efforts. This is one reason the Essence framework provides the pattern element supporting an organization's extension of the framework.

Patterns can help you transfer knowledge faster from your experts to your less experienced personnel helping you grow the competencies you need faster by sharing the critical knowledge of your experts.

This is a case that demonstrates under the stress of a real project, or due to inadequate experience, fundamentals are easily forgotten. Even experienced managers sometimes let project specific conditions blur their vision. The Essence thinking framework helps us cut through the day to day project confusion getting right to the core issues that can help us make the best decisions.

# Pattern 6: Taking Responsibility for a Key Requirement

*Often the issues faced even on complex efforts return to fundamentals as we saw in the previous example. We observed another example of missed fundamentals earlier in the book in the story where no one wanted to own a key requirement because everyone was too busy. This example should be of particular interest to all team members.*

This pattern is based on the story from Chapter Four about the small change requiring practice to master. Because all the project leaders had more work on their plate than they could handle, no one wanted to take on more responsibility to own a key requirement for a support system. As a result everyone just began ignoring a major risk that should have been raised and mitigated early. This story relates to the Work Alpha. Refer to Figure P6-1.

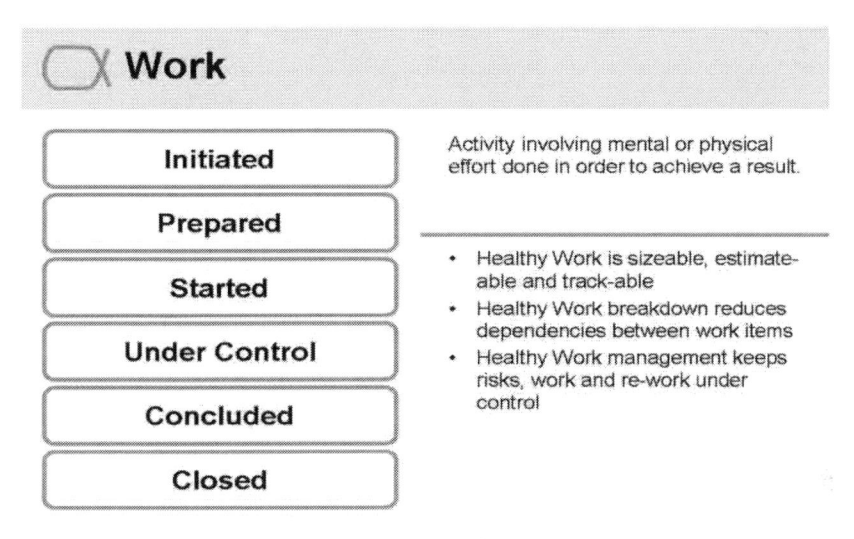

Figure P6-1 Work Alpha

The *Work* Alpha is defined as:

- *Activity involving mental or physical effort done in order to achieve a result.*

The relevant state in this story is the *Under Control* state. Refer to Figure P6-2.

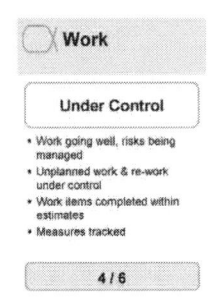

Figure P6-2 Work Alpha Under Control State

If the team had been using the Essence framework to assess the progress and health of this project, they should have known that they had not achieved the work *Under Control* state. This state is defined as:

- *The work is going well, risks are under control and productivity levels are sufficient to achieve a satisfactory result.*

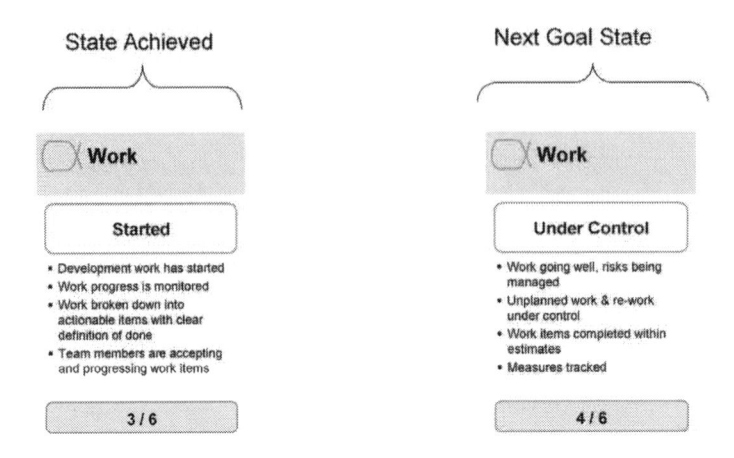

Refer to Figure P6-3.

**Figure P6-3 Work Alpha Progress and Health Assessment**

Recognizing that all of the required work was not under control because a key part of the work wasn't getting done should have led the team to question whether they had the appropriate management and leadership competencies on the team. At a minimum, if there was no other clear action they felt they could take, someone should have raised a risk and the risk should have been maintained on the risk list until action was taken to mitigate it, or until it was no longer a risk. Risk is not just associated with the Work Alpha.

When a group of software professionals in a large US defense company were asked if they could see the value of the Essence framework after the framework idea had been explained to them, one experienced software professional commented that they saw the complete framework as a potential aid to implement an effective risk management practice.

Effective risk management is essential to all software endeavors, but risks are not limited to any single Alpha. The risk of a software endeavor is affected by the health of all the Alphas.

Refer to Figure P6-4 for a diagram of this pattern. The entry criteria to trigger this pattern is the recognition that no one on the project team is owning a key requirement.

**Pattern Trigger:**
No one on project team owing a key requirement

**Taking responsibility for a key requirement pattern**

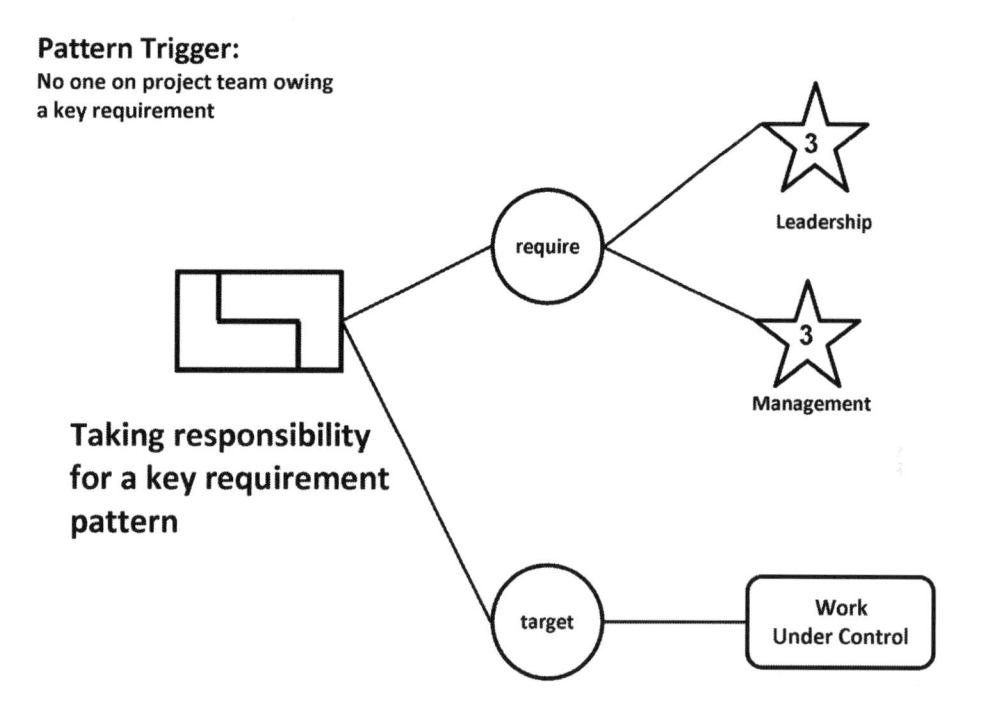

Figure P6-4 Taking Responsibility for a Key Requirement Pattern

It is also worth noting that there isn't just one way to detect a potential problem (e.g. a risk) when using the Alphas. In the case of the *Taking responsibility for a key requirement pattern* one could argue that the root cause should have been detectable early in the project when assessing the Team Alpha. Two of the states of the Team Alpha are Team Seeded and Team Collaborating. Two of the checklists items for the Team Seeded state are:

- *The team's responsibilities are outlined*
- *The level of team commitment is clear*

And two of the checklist items for Team Collaborating state are:

- *The team is working as one cohesive unit*
- *Communication within the team is open and honest*

When assessing the Team state it is likely the team would have discovered that either there was a problem in the way the team's responsibilities had been defined, or there was at least a problem in the level of team commitment.

If this had not become evident when assessing the Team Seeded state, then later when assessing the Team Collaborating state, the fact that this risk was continuing to be ignored should have caused some of the team members to question whether the team was in fact working as one cohesive unit and whether they were communicating within the team in an open and honest fashion.

# Pattern 7: Using Data to Aid Practical Improvement

*The next example isn't one I was personally involved with, but one that Jeff Sutherland, who taught me Scrum fundamentals, has written extensively about.*[59] *[50] This example demonstrates fundamentals that can work well with both high maturity organizations and agile organizations and it demonstrates how the checklist mechanism, discussed throughout this book, can be a great aid to sustainable high performance. This example should be of particular interest to organizations using the CMMI or any other performance improvement model (e.g. lean six sigma, agile retrospectives) and are looking for practical ways to increase agility and performance.*

What too few organizations understand is that true high maturity means continually learning from your data better ways to prevent problems before they occur and sharing those lessons across your organization. I now want to describe an example where a company used CMMI high maturity practices (level 4 and 5) together with Scrum and checklists to improve their performance and shared those improvements across their organization.

Systematic, a company in Denmark, used Scrum and CMMI Level 4 and 5 practices to achieve:

- *"sustainable quadrupling of productivity compared to waterfall projects".*[50]

---

[59]"Scrum and CMMI: Going from Good to Great" by Jeff Sutherland and Carsten Jakobsen

One of the measures they used on the project was "story point efficiency" which was defined as the ratio of their estimated time to complete the implementation of a backlog story to the actual time it took. The team chose this measure and set a goal to improve it because they recognized that the work that was prioritized for upcoming sprints was not being sufficiently prepared in a timely manner.

The problem Systematic was encountering was not unlike what I described earlier in the Pattern, Reasoning Through a Requirements Dilemma. When requirements in the backlog are not prepared properly it results in project delays and wasted effort. As is often the case, when teams look close, as the Systematic team did, at what is causing these delays they often can identify some very specific things that, if worked earlier, could help avoid the delays.

In this example, the team learned to recognize common *patterns* leading to low story point efficiency including *vague customer requirements* not sufficiently understood, *risks* that had not been appropriately identified and mitigated, stories that had *not been properly broken down and estimated*, and *insufficient technical design*.

To improve their story point efficiency the team learned to recognized these patterns earlier in each sprint, and reject backlog items that had not been properly prepared sending them back to the product owner. They also produced a *checklist* to help the product owner resolve these issues earlier so fewer backlog items would be rejected in the future.

The checklist items included:

- *Customer requirements sufficiently understood*
- *Risks identified*
- *Stories estimated*
- *Technical design drafted (no uncertainties)*

Notice how these checklist items align with Essence framework checklist items. Requirements Alpha: *Coherent* State checklist item:

- *The impact of implementing the requirements is understood.*

Refer to Figure P7-1.

**Requirements**

**Coherent**

- Described requirements provide coherent picture of the system
- Conflicting requirements separated
- **Impact of implementing requirements understood** ← **Customer requirements sufficiently understood**
- Priority of requirements clear

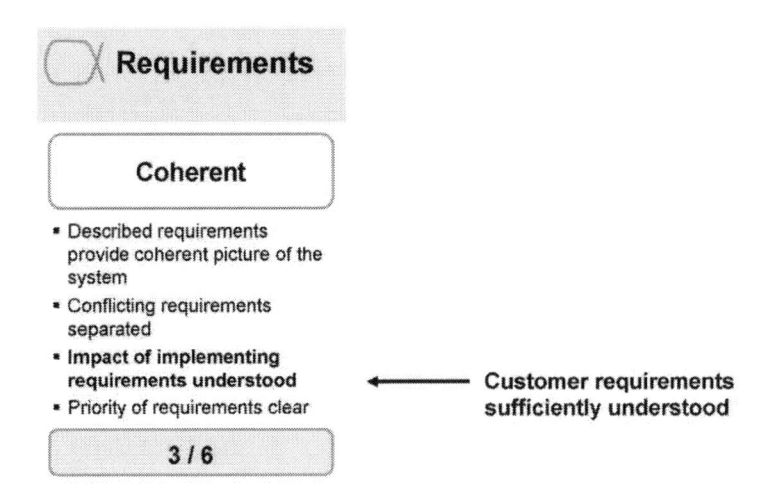

3 / 6

Figure P7-1 Requirements Alpha Coherent State Correlation

Work Alpha: *Prepared* state checklist items:

- *Risk exposure is understood*
- *Cost and effort of the work are estimated*

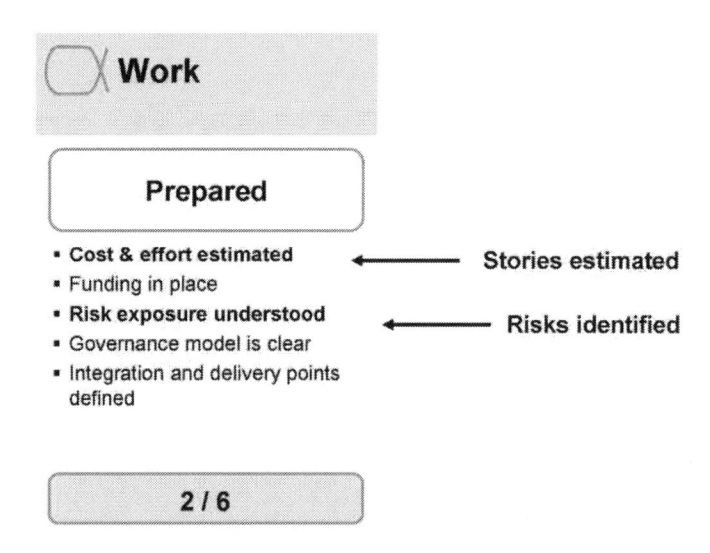

Refer to Figure P7-2.

**Figure P7-2** Work Alpha Prepared State Correlation

Software System Alpha: *Architecture Selected* state checklist item:

- *Architecture selected that addresses key technical risks*

Refer to Figure P7-3.

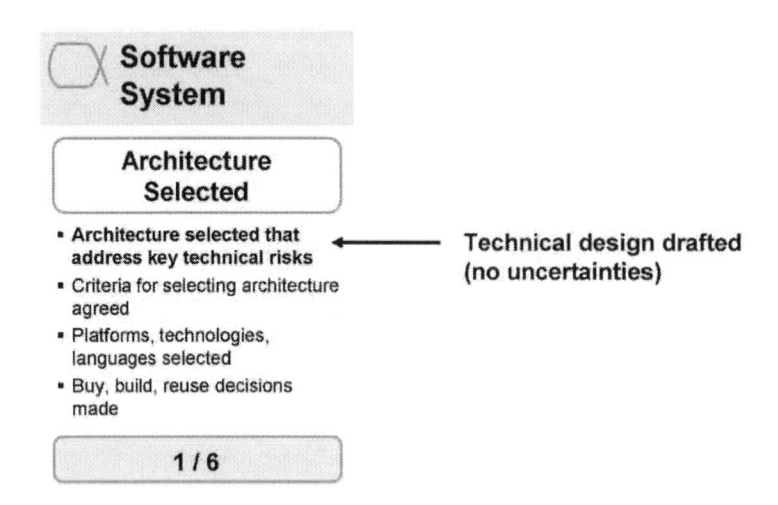

```
┌─────────────────┐
│ ◯╲  Software    │
│ ◯╱  System      │
└─────────────────┘
    ┌──────────────┐
    │ Architecture │
    │   Selected   │
    └──────────────┘
• Architecture selected that    ◄──────  Technical design drafted
  address key technical risks            (no uncertainties)
• Criteria for selecting architecture
  agreed
• Platforms, technologies,
  languages selected
• Buy, build, reuse decisions
  made

    ┌──────────────┐
    │     1 / 6    │
    └──────────────┘
```

Figure P7-3 Software System Alpha Architecture Selected State Correlation

Also note how these checklist items are focused on *prevention* of common roadblocks observed in the past, such as by helping the product owner address product backlog readiness issues in a timely way.

Similarly, recall in my personal improvement story earlier in the book the way I improved sustaining my performance was by recognizing common situations ahead of time where I could take deliberate actions early thereby avoiding those patterns that often would lead to trouble.

These are examples that demonstrate the use of data in a *predictive* way. This means using your data to predict what is likely to occur and taking action that increases the likelihood of avoiding problematic situations.

Checklists, used in this way, are a preventive aid. Preventing problems before they occur is the best route to sustained high performance. This is a practical example of how effective, or true high maturity organizations operate. Refer to Figure P7-4 for a diagram of this pattern.

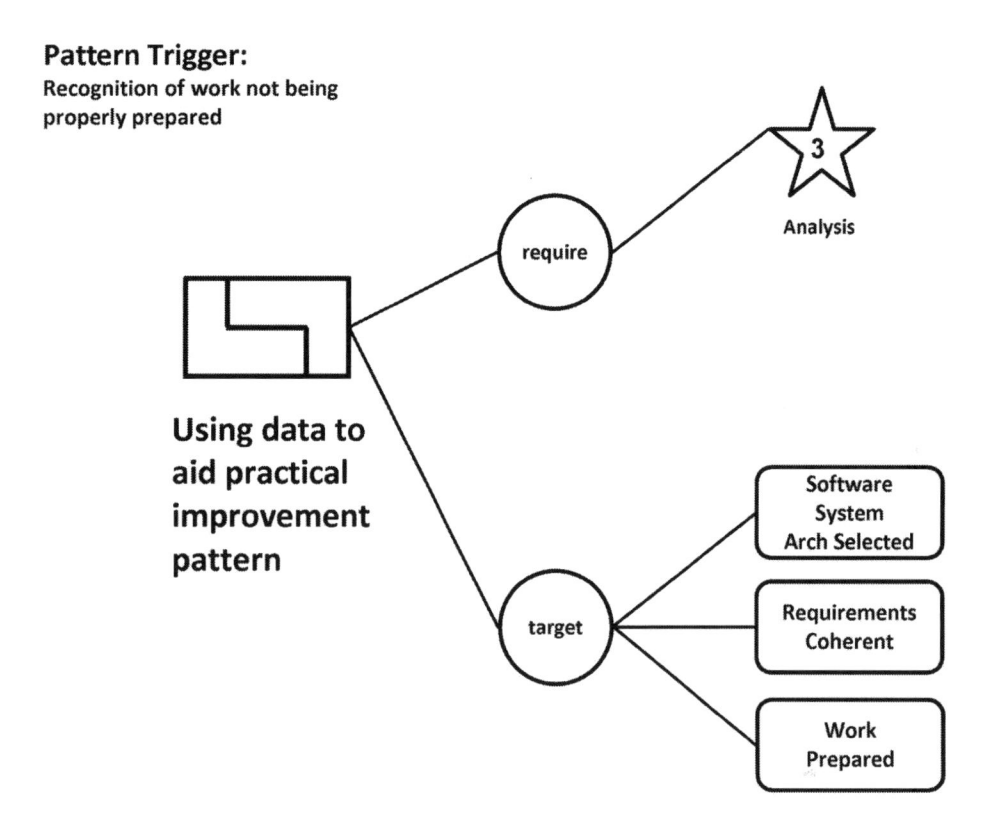

Figure P7-4 Using Data to Aid Practical Improvement Pattern

The entry criteria, or trigger, to apply this pattern is recognition of work not being properly prepared. It is also worth noting that this practical high maturity pattern demonstrates how the Essence framework can be used in a complementary way to both agile approaches (e.g. Scrum) and other industry improvement models (e.g. CMMI).

# Pattern 8: The Late Hardware Dilemma

*Many of the opportunities for improvement come back to simply recognizing specific situations that arise and capturing those situations as common patterns as we saw in the last example. We now revisit another example discussed earlier in the book that demonstrates the power of pattern recognition to aid sustained high performance. This example should be of particular interest to all team members.*

Recall in the procurement department case discussed earlier in the book that one of the reasons for the frequent late hardware turned out to be the fact that the procurement personnel were often over worked and stressed and under these conditions frequently couldn't recall what to do when faced with a specific situation related to missing information on a procurement requisition form. By capturing just a few patterns we were able to help them recognize the most common situations, and make more accurate and rapid decisions. This helped procurement personnel manage their work and keep it under control. We effectively helped them raise their work management competency faster than they could have done it without this assistance.

This example relates to the *Management* competency, the *Work* Alpha and the *Under Control* state within the Work Alpha. When using the Essence framework you have multiple alternatives to communicating useful patterns to your practitioners. You can express the pattern graphically using the pattern construct in the language and use such examples in training classes. Refer to Figure P8-1.

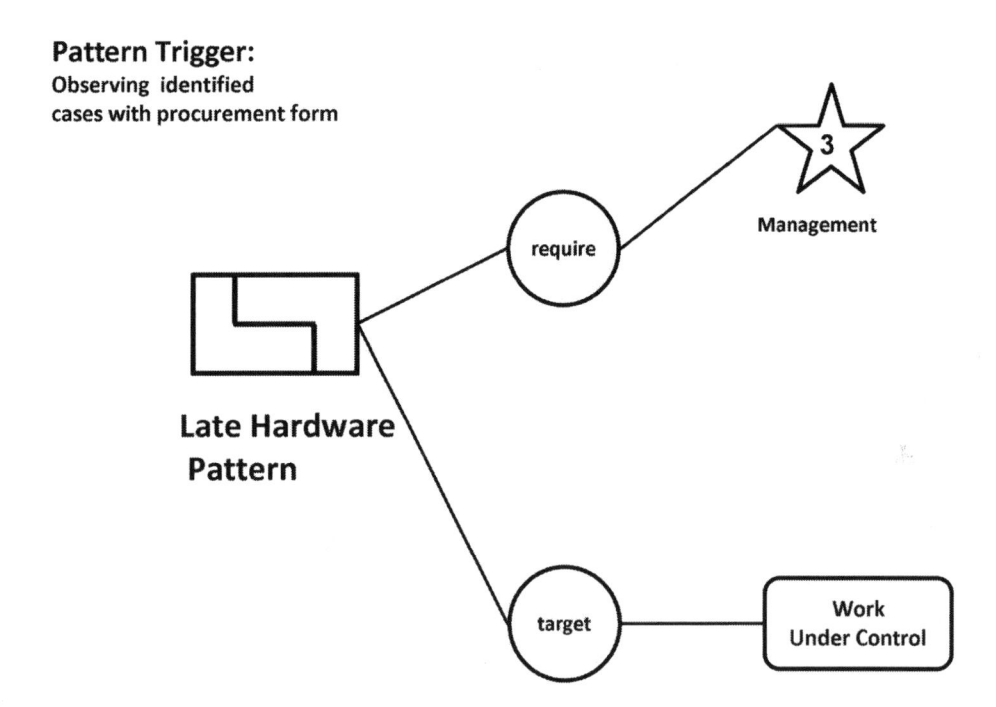

Figure P8-1 Late Hardware Pattern

You could also add a hint as a reminder to a state card. Refer to Figure P8-2.

## Work

**Under Control**

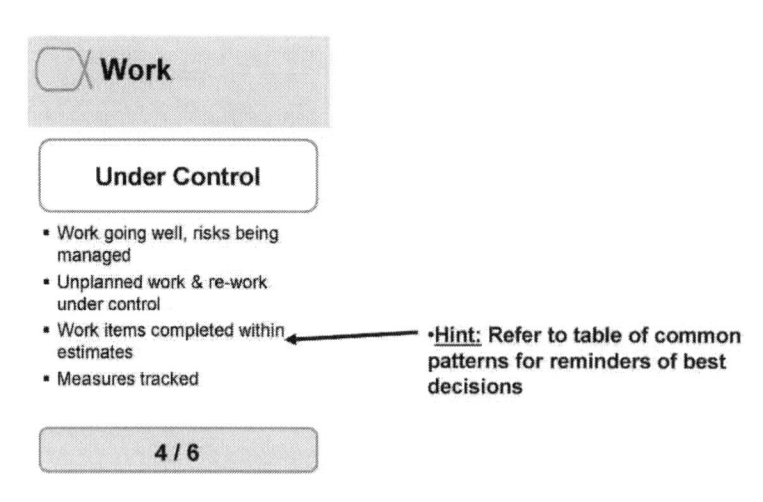

- Work going well, risks being managed
- Unplanned work & re-work under control
- Work items completed within estimates
- Measures tracked

•**Hint:** Refer to table of common patterns for reminders of best decisions

4 / 6

Figure P8-2 Reminder Hint on State Card

You could also use the state cards to communicate the state (s) achieved and next goal state (s) as we have seen previously in progress and health assessment figures using the state cards. Another option would be to add an additional checklist item.

# Pattern 9: Assisting Self-Direction

*In Chapter 14 we discussed a simple pattern referred to as "Gentle Reminders." But sometimes the situation is more difficult and gentle reminders are not sufficient. This is often the case when a team is trying to change, but they are missing critical competencies. In this case practice with gentle, or soft, reminders may not be sufficient. This example should be of particular interest to organizational leaders who are interested in knowing whether the competencies of their people are sufficient to meet current and future organizational objectives.*

We previously discussed the Team Alpha in Chapter 13 and 14. As you listen to this scenario think about the Team Alpha again, only this time think about the Team Formed state and the following checklist item:

- *Team members understand their responsibilities and how they align with their competencies*

Refer to Figure P9-1 for the Team Alpha Formed state card.

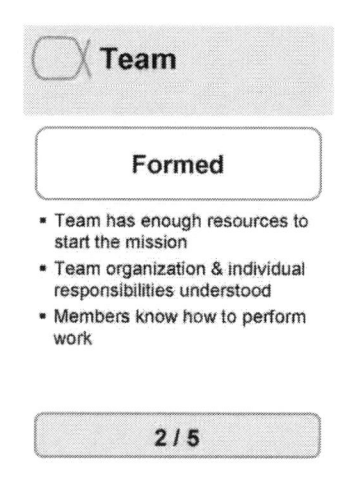

**Figure P9-1 Team Alpha Formed State**

- Team has enough resources to start the mission
- Team organization & individual responsibilities understood
- Members know how to perform work

(2 / 5)

In 2011, despite the popularity of agile approaches, significant disillusionment with agile methods was evident across a large portion of the software community. An article appearing in June, 2011[60] [11] attributed at least part of this disillusionment to three factors:

a. Too little analysis being done to address agile's appropriateness to an organization's situation.
b. Trying to force a strict agile approach when it is not appropriate for every environment.
c. Misunderstanding agile principles as a replacement for discipline.

In this example I share some details from one disillusioned organization which I worked with that demonstrates the significance of the issues still being faced.

This was a traditionally structured organization with a hierarchical command and control leadership style (still common in many organizations today). This organization decided it needed to transform itself into an agile organization to remain competitive. So the command came down through the management chain:

[60]http://searchsoftwarequality.techtarget.com/feature/Agile-development-Whats-behind-the-backlash-against-Agile?

*"We are now agile so managers need to stop telling the software teams what to do!"*

It took less than six months before many project teams were in deep trouble. Management wanted answers. I was asked to conduct an independent evaluation, and it didn't take long to figure out what was going on. The first person I talked to was a member of a software team. I asked him how the move to agile was going. He replied:

*"Great! Management is finally off our backs."*

When I asked him if his team was meeting their schedule and cost commitments he said he couldn't say because he wasn't a manager. The executives in this organization had commanded:

*"We are now agile!"*

But no one told the software teams how that affected their responsibilities.

Everyone in this organization seemed to think that being agile simply meant the project managers would stop bothering the software teams. What they didn't realize was that agile teams are self-directed teams, which means the responsibilities for planning, estimating, tasking, measuring, collaborating with stakeholders and addressing risks now fall inside the software teams.

The value of the Essence thinking framework is that it can be used in cases such as this to point out gaps in the team's practices and the team member's competencies that the team may not see themselves. In this example the team was unaware that their responsibilities had changed given the change to the new agile approach, and they were unaware of the new competencies they needed in order to achieve and sustain high performance.

In this case, if the team had been using the Essence thinking framework when they assessed the Team Alpha Formed state at least some of the team members should have recognized that they had not achieved the following checklist items:

- *Team members understand their responsibilities and how they align with their competencies*
- *All team members know how to perform their work*

Refer to Figure P9-2 for a diagram that shows how this team could have assessed the Team Alpha.

**Figure P9-2 Progress and Health Assessment**

The team may have decided they had achieved the Team Seeded state, even if some of the team members did not know what it meant to be agile, as long as at least a few of the team members raised the issue of the need for additional competencies with respect to self-direction skills. But when they assessed the Team Formed state they should have recognized that many of the team members did not know how to perform their work given the new responsibilities of self-direction that comes with an agile way of working.

Refer to Figure P9-3 for a diagram capturing the Assisting Self-Direction Pattern.

**Pattern Trigger:**
Change in practice expectations

**Assisting Self-Direction Pattern**

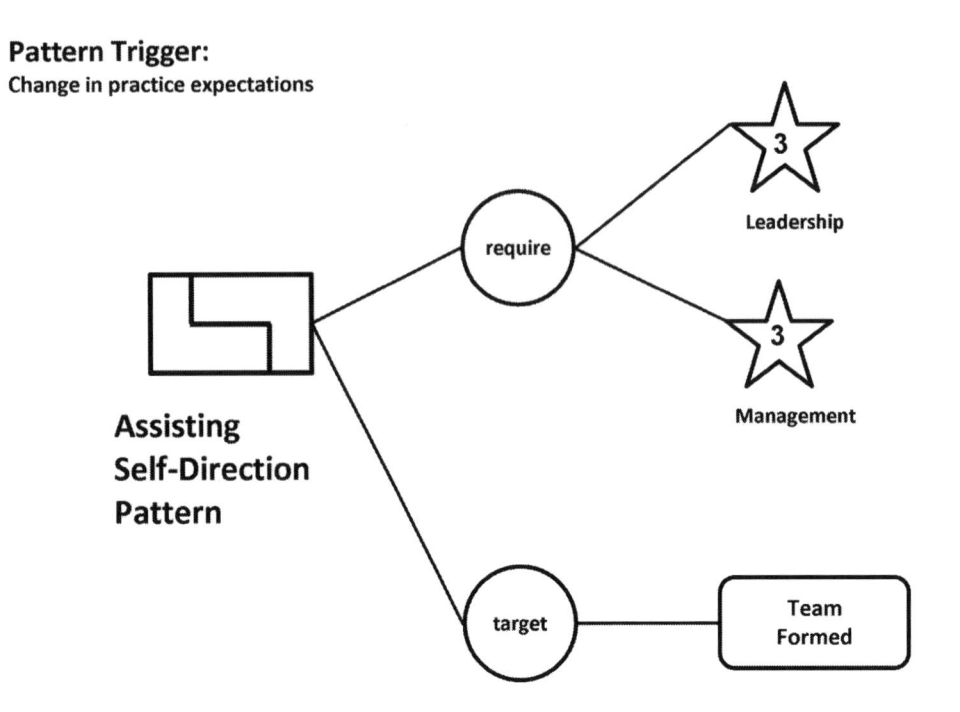

Figure P9-3 Assisting Self Direction Pattern

This pattern should be triggered when team members recognize that a change in expected practices has occurred and they are unsure of their new responsibilities.

# Appendix Q – Practical High Maturity and Essence

*In Part III and Appendix P we saw examples of using the Essence Thinking Framework to improve our performance by reminding us of patterns that can help us assess where we are more effectively and make better on the job decisions. Nevertheless, sometimes even when people want to make a change to improve performance, they may not understand the full consequences of the change they are taking on. In these cases you may need more than just gentle reminders or training. You may need to use your data to impact a critical decision. This example should be of particular interest to project leaders, managers and organizational leaders interested in practical high maturity.*

In business a common situation that occurs is where a team is asked to *"take a challenge"* accepting an aggressive project schedule without adequate analysis of the workload. This situation all too often is used by management with the thought that it will just motivate the team to work harder. However, too often it leads to costly inappropriate decisions once the team realizes good options have been exhausted. Lets now briefly review what we discussed earlier in the book about how our brain works and how it affects our performance before we investigate this case further.

### Reviewing How the Brain Works and How We Can Counter its Negative Effects

While most of us try to use our brain to find the truth, the way our brain actually works—particularly in stressful situations–can actually mask the truth from us. To understand this apparent paradox we just need to revisit what we have already discussed about how the two sides of our brain operate—and in particular how each side responds differently under stress.

First, when we are feeling stressed, which is not uncommon when reporting status up the chain in business, or talking to a coach, or communicating to a customer or other key stakeholder, often the emotional side of the brain takes charge. As discussed earlier in the book the emotional side of our brain does not like uncertainty [61] and

---

[61] Jonah Lehrer explains the drive of the emotional brain to seek patterns in his book *"How We Decide"*.

it will desperately seek to avoid uncertainty even if that means seeking out false patterns.

As living creatures, survival is often foremost on our mind. We would rather build a story that fits what we think others want to hear than put our personal survival at risk. Unfortunately, this can lead to poor decisions with respect to long term performance goals, professional or personal. In the common business case "*take a challenge*" the false pattern the emotional brain jumps to is:

*"Yes, we can do it, if we just keep a positive attitude!"*

Why do we jump to this common false pattern? The answer is because the emotional side of the brain is constantly telling us not to bring up information that might raise uncertainty. But this is where the rational side of our brain can help. The rational side can help our performance by bringing up risks early so they can be dealt with in a timely way ultimately reducing risk to our desired long term sustained performance. But to do this we need to engage the rational brain at just the right time to ensure the emotional brain is not ignoring important objective and contextual data that might be indicating a different decision should be made right now.

To sustain higher levels of performance we must constantly battle the emotional brain's desire to jump to the first pattern it sees which too often is a false pattern, or not the most effective pattern given the current situation. Once we understand how our brain works we can then begin to institute the right practice scenarios to help us get better at engaging both sides of the brain at just the right time leading to better decisions and more sustainable performance.

So how do we counter common false patterns? The answer can be found in what we learned earlier in the book about how Tom Brady gets ready for his next big game. It's called practice, but its not any form of practice. Rather it is practice with the right objective and contextual data conducted at just the right time.

As I explain below how I help people with this situation in my workshops think about the *Team* and *Work* Alphas in the Essence Thinking Framework. Refer to Figure Q-1.

## Team

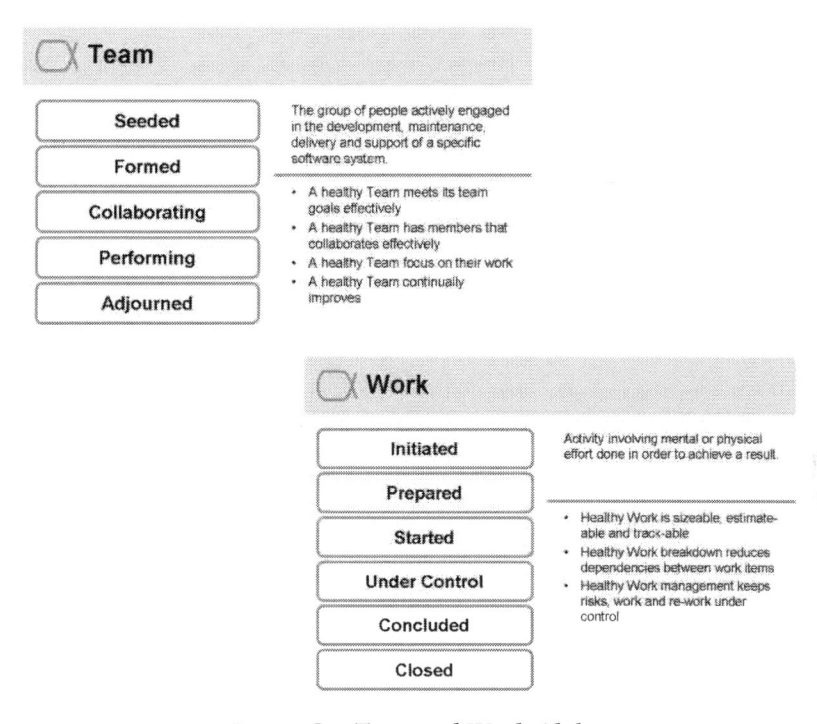

- Seeded
- Formed
- Collaborating
- Performing
- Adjourned

The group of people actively engaged in the development, maintenance, delivery and support of a specific software system.

- A healthy Team meets its team goals effectively
- A healthy Team has members that collaborates effectively
- A healthy Team focus on their work
- A healthy Team continually improves

## Work

- Initiated
- Prepared
- Started
- Under Control
- Concluded
- Closed

Activity involving mental or physical effort done in order to achieve a result.

- Healthy Work is sizeable, estimateable and track-able
- Healthy Work breakdown reduces dependencies between work items
- Healthy Work management keeps risks, work and re-work under control

**Figure Q-1 Team and Work Alphas**

Particularly, think about the Team *Collaborating* state and the Work *Under Control* state. Refer to Figure Q-2

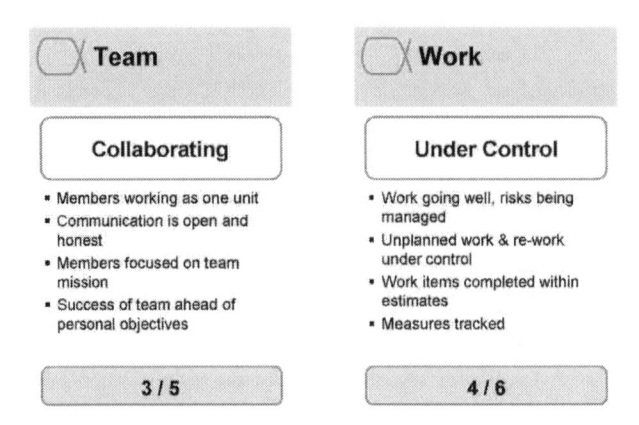

## Figure Q-2 Team Collaborating State and Work Under Control State

**Team — Collaborating** (3/5)
- Members working as one unit
- Communication is open and honest
- Members focused on team mission
- Success of team ahead of personal objectives

**Work — Under Control** (4/6)
- Work going well, risks being managed
- Unplanned work & re-work under control
- Work items completed within estimates
- Measures tracked

The Team *Collaborating* state is defined as:

- *the team members are working together as one unit*

One of the checklist items in this state is:

- *Communication within the team is open and honest*

The Work *Under Control* state is defined as:

- *the work is going well, risks are under control, and productivity levels are sufficient to achieve a satisfactory result*

Two of the checklist items in this state are:

- *Estimates are revised to reflect the team's performance.*
- *Measures are available to show progress and velocity*

### Practicing Your Practices To Counter Common False Patterns

In my workshops to help people learn integrated practice techniques, I teach them how to prepare for an upcoming discussion by keeping both parts of their brain appropriately engaged at just the right time. This is where they learn to build a *Structured Real Story* (refer back to Chapter Ten for discussion of Structured Real Stories).

The way you go about building the story is key because it engages both sides of your brain helping you first logically figure out the right story yourself, as you build it. It is not uncommon during these exercises for people to discover that the story they thought they were going to build turns out to have a different ending from what they originally believed. This is why taking the time to build the story and think it through in a logical fashion is so important.

As an example, one of the scenarios I give as a workshop exercise is what I refer to as the *"Team Player Scenario"*. In this scenario I present a situation to the class participants where the group plays the role of a project engineer on a subcontract to a larger organization. In this scenario a representative from the prime contractor asks them to shorten their schedule by five weeks. I usually present this scenario after we have discussed issues related to subcontract management, and the importance of prime contractors and subcontractors working together as a single collaborating team.

I then ask the participants to tell me how they would respond to such a request. I usually get plenty of different approaches, but it is not uncommon for someone to respond with something like:

*"I would tell them that we are team players. I know we can do it!"*

Often the class thinks this might be the right answer and they usually are supportive of this response thinking it is what I am looking for. This is because we have just finished discussing the importance of being team players and working closely with your prime contractor. But this is where I throw them a curve. In this scenario I play the role of senior management and I surprise the class when I respond:

*"How do you know you can do it? Where is your data? Show me, don't tell me."*

Then I ask the participants if anyone can think of a better option? If we have already discussed the *Structured Real Story* technique some of them may bring it up here. If not, I might explain how the Structured Real Story technique can help you in this kind of situation. By gathering objective and contextual data[62] and then playing devils advocate by asking yourself the right questions ahead of time you can be prepared to answer this type of question from a Senior Manager or a teammate with something like:

*"The prime contractor wants us to reduce the schedule by five weeks. We have assessed our work and resources. We can take two weeks out of the schedule if we can keep John on the project for one additional month. We have assessed our current list of tasks remaining and assignments. We have the data if you would like to review it. We don't see how we can shorten the schedule any more without adding unacceptable risk."*

**Structured Real Story Technique Data for Team Player Scenario**

*Objective Data: Prime wants schedule reduced by 5 weeks*

*Contextual Data: Assessed situation including list of tasks remaining*

*Consequence: If reduce schedule more than 2 weeks, result is unacceptable risk*

*Recommendation: Can take 2 weeks out of schedule if keep John on project for additional month*

**An Example of Practical High Maturity Using the Essence Thinking Framework**

What you learn when you use the Structured Real Story technique is how to take the emotion out of the discussion. By building the story using both sides of your brain you have already begun your *integrated practice* (or practicing your practices) helping you prepare to present your case. By showing the data you used to arrive at your conclusion you can draw a line in the sand making it easier for you to stick to the decision you know is right. This is another example, as we discussed in Chapter Ten, of Practical High Maturity, or using real data to predict and make better decisions today.

From the Essence thinking framework perspective what we are doing in this workshop is helping people learn one way to get from where they are to where they need to go next on their endeavor to be successful. More specifically, this

---

[62] Refer to Chapter Ten for more information about using high maturity techniques in a practical way to help practitioners gather objective and contextual data. This is an area where the traditional process engineer's role can evolve when using the SEMAT approach to better support practitioners on the job daily needs.

technique can help a team achieve the Work Alpha Under Control State and the Team Collaboration state. Refer to Figure Q-3.

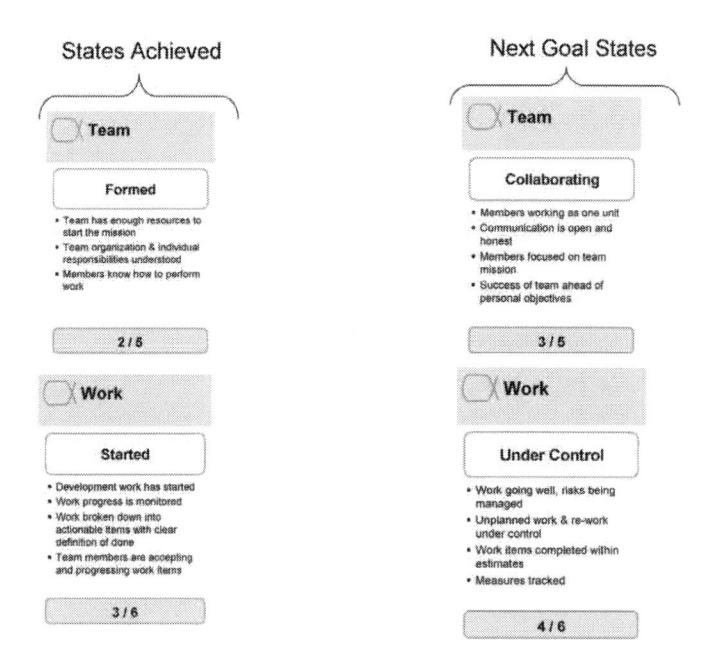

**States Achieved**

- Team — Formed
  - Team has enough resources to start the mission
  - Team organization & individual responsibilities understood
  - Members know how to perform work
  - 2 / 6

- Work — Started
  - Development work has started
  - Work progress is monitored
  - Work broken down into actionable items with clear definition of done
  - Team members are accepting and progressing work items
  - 3 / 6

**Next Goal States**

- Team — Collaborating
  - Members working as one unit
  - Communication is open and honest
  - Members focused on team mission
  - Success of team ahead of personal objectives
  - 3 / 5

- Work — Under Control
  - Work going well, risks being managed
  - Unplanned work & re-work under control
  - Work items completed within estimates
  - Measures tracked
  - 4 / 6

**Figure Q-3 Achieving Team Collaborating State and Work Under Control State**

The workshop participants are now prepared to continue their integrated practice by running through the story over and over like Tom Brady runs through the game tapes of his next opponent. They learn to think-through different possibilities and as they do they are mentally preparing for each potential situation that might arise. In the workshop I often ask the participants to think about other situations that might arise, and other questions a senior manager or stakeholder might ask in this situation such as:

*"What if I give you John for two additional months, or what if we got the customer to agree to drop that really tough requirement we talked about last week?"*

> **Sidebar: Not just for managers**
>
> The Structured Real Story technique is not just for managers and for reporting "up the chain". It can help a worker communicate with any stakeholder. Communication skills are needed by all members of self-directed teams and are particularly important to help people achieve higher leadership and management competencies faster.

The point is they learn the importance of thinking through ahead of time the situations that most often occur so they can respond in a prepared way with real data—not just emotion. They also learn in the workshop that they can't just leave these techniques in the workshop. They learn what they need to do after they leave the workshop and go back to their real job. They have to take the process back and use it refreshing the real data that keeps changing every day in their real project environment. This helps them prepare not just for senior manager briefs, but for collaborative interactions with their teammates, and the most likely scenarios and decisions they will face each day on the job.

**Relationship of Structured Real Story Technique to Essence Thinking Framework**

From the Essence thinking framework perspective the technique described of using objective and contextual data to build a Structured Real Story and then practice your story by thinking through the issues is just one way to help you achieve the Team Collaborating state and the Work Under Control state. Specifically it helps you achieve the checklist item:

- *Communication within the team is open and honest*

And it helps you achieve the following two checklist items:

- *Estimates are revised to reflect the team's performance*
- *Measures are available to show progress and velocity*

Refer to Figure Q-4.[63]

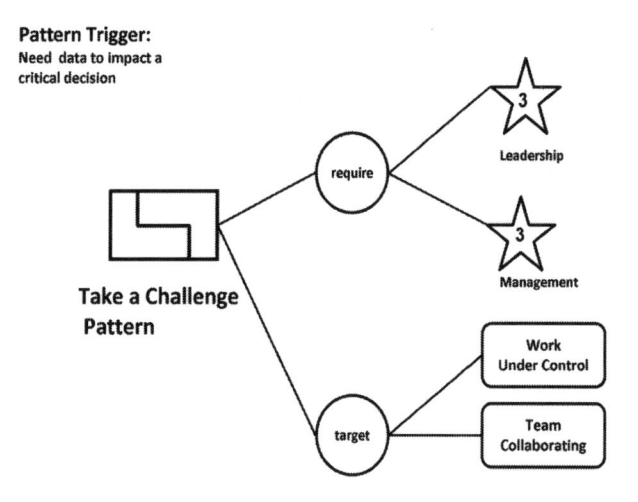

**Figure Q-4 Take a Challenge Pattern**

The "Take a Challenge" pattern could be triggered into action when a team recognizes it needs objective and contextual data to impact a critical decision.

The practice of building a Structured Real Story is not required with the Essence thinking framework. As mentioned earlier the framework includes no practices, just the essentials. You decide what practices, patterns or techniques you need to help your team perform and sustain their performance.

---

[63] The "Take a Challenge Pattern" could also be referred to as the "Team Player Scenario". The essentials of both scenarios relate to Team Collaborating, Work Under Control and the Management and Leadership competencies.

**Appendix Q Summary Key Points**

- Always keep clear in your mind the fact that the emotional side of our brain does not like uncertainty and it will desperately seek to avoid uncertainty even if that means seeking out false patterns.
- To sustain higher levels of performance we must constantly battle the emotional side of the brain's desire to jump to the first pattern it sees which too often is a false pattern, or not the most effective pattern given the current situation.
- Counter common false patterns with practice, but not just any form of practice. It must be practice with the right objective and contextual data conducted at just the right time.
- The Essence thinking framework includes no practices, just the essentials. You decide what practices or techniques you need to help your team perform and sustain their performance.

# Appendix R – Helping Your Team Achieve Higher Competency Faster

The Essence thinking framework includes six competencies required to perform software engineering, each having five competency levels. This framework includes essentials for software engineering, but an organization could decide to extend the framework to address the needs of other disciplines– such as a procurement department. Let's now take a closer look at the Essence framework competencies and the competency levels. Refer to Figure R-1 for a diagram identifying the Six Essence Framework competencies.

# Appendix R – Helping Your Team Achieve Higher Competency Faster

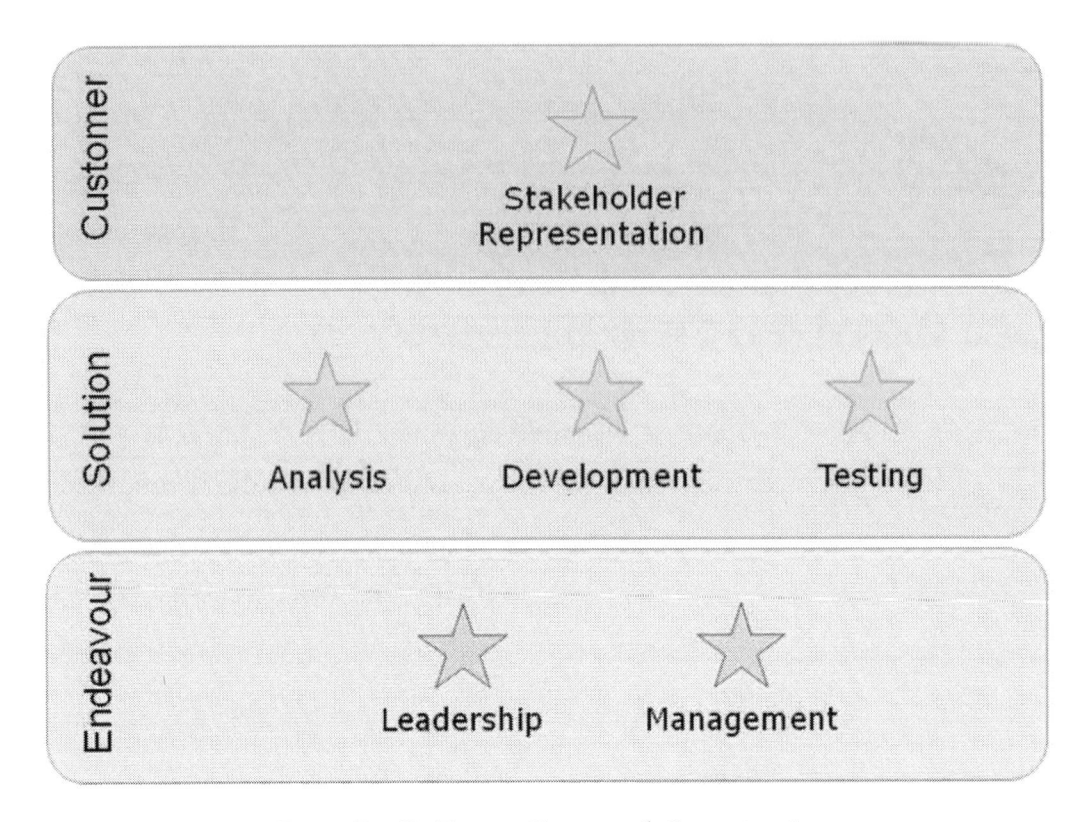

**Figure R-1 Six Essence Framework Competencies**

Following is a summary of the definitions of the six Essence framework competencies:

Stakeholder Representation: This competency encapsulates the ability to gather, communicate and balance the needs of other stakeholders, and accurately represent their views.

Analysis: This competency encapsulates the ability to understand opportunities and their related stakeholder needs, and transform them into an agreed and consistent set of requirements.

Development: This competency encapsulates the ability to design and program effective software systems following the standards and norms agreed by the team.

Testing: This competency encapsulates the ability to test a system, verifying that it is usable and that it meets the requirements.

Leadership: This competency enables a person to inspire and motivate a group of people to achieve a successful conclusion to their work and to meet their objectives.

Management: This competency encapsulates the ability to coordinate, plan and track the work done by a team.

Following is the definition for the five Essence competency levels:

1. Assists: Demonstrates a basic understanding of the concepts involved and can follow instructions.
2. Applies: Able to apply the concepts in simple contexts by routinely applying the experience gained so far.
3. Masters: Able to apply the concepts in most contexts and has the experience to work without supervision.
4. Adapts: Able to apply judgment on when and how to apply the concepts to more complex contexts. Can enable others to apply the concepts.
5. Innovates: A recognized expert, able to extend the concepts to new contexts and inspire others.

In the case studies in this book you have seen examples that require a minimum of the level 3 (Master level) for the Stakeholder Representation, Analysis, Leadership and Management competencies.

The majority of organizations I have observed spend more time defining roles and responsibilities than identifying the needed competencies and competency levels required to perform those roles. However, assigning a role with a list of responsibilities does not ensure people are competent to perform that role, especially under the varying situations they are likely to face each day on the job.

As discussed in the first chapter of this book many organizations have supplemented their roles and responsibilities with classroom training, but this type of training is often inadequate when it comes to ensuring people know how to perform well under the varying situations they often face on the job.

The Essence thinking framework can help all practitioners, including students and less experienced practitioners, achieve higher competency faster. Sharing with your people the case study examples in this book and similar patterns more reflective of the situations your practitioners face most often can help any organization improve the competency of their people and their overall organizational performance.

One of the primary reasons the Essence approach can help you move your organization's competency level to where it needs to be faster than traditional approaches is because it provides greater transparency into what your team is actually doing and the competency requirements to accomplish the work at hand.

To gain this transparency requires open and honest communication which can only be achieved in an environment where trust exists. The Essence approach supports this need by providing a safe mechanism to put your team in charge of the way they work, and their own self-assessment and continual improvement. It also provides a mechanism for your team to share more rapidly with less experienced team members the knowledge needed to perform effectively as contributing team members.

Today project conditions are changing faster and faster and success requires faster responses which are best achieved through those with the most relevant and specific knowledge. But success also requires that those making critical decisions have the required competencies. By giving your practitioners the responsibility to improve their own practices you also give them the responsibility to grow their own competency which is essential for sustained higher organizational performance.

# Appendix S – Cross Reference Critical Competency Needs, Case Studies

In this appendix I identify fifteen case studies/patterns discussed in this book along with critical competencies associated with each. I also provide a cross-reference to help you locate each in the book (in parenthesis after the Case Study/Pattern). Refer to the Table below. Following the referenced table you will find observations about these case studies based on my personal experiences as a software practitioner, manager, leader, coach and consultant spanning the past forty years.

| Case Study/Pattern (Chap Ref) | Critical Compentencies |
|---|---|
| 1 Common Measurement Mistake(Ch 3,10) | Leadership, Management, Stakeholder Rep. |
| 2 Business Case Study Sustained Performance (Ch 4) | Leadership, Management |
| 3 "Big Picture Feel" Planning Case (Ch 6) | Leadership, Management |
| 4 Utilizing "Listen for" Tecnique for Positive Outcomes(Ch 9) | Leadership, Management, Stakeholder Rep. |
| 5 Dictated Schedule(Ch 10,14) | Leadership, Management, Stakeholder Rep. |
| 6 Example Structured Real Story Technique(Ch 10) | Management, Stakeholder Rep. |
| 7 Pattern: Keeping an Opportunity Alive (Ch 14) | Leadership, Stakeholder Rep. |
| 8 Pattern 4: Reasoning Through a Requirements Dilemma(App P) | Analysis, Stakeholder Rep. |
| 9 Pattern 5: Reasoning Through a Lack of Data Challenge(App P) | Leadership, Management, Stakeholder Rep. |
| 10 Pattern 6: Taking Responsibility for Key Requirement(App P) | Leadership, Management, Stakeholder Rep. |

# Appendix S – Cross Reference Critical Competency Needs, Case Studies

| Case Study/Pattern (Chap Ref) | Critical Compentencies |
|---|---|
| 11 Pattern 7: Using Data to Aid Practical Improvement(App P) | Analysis |
| 12 Procurement Case Case (Ch 2, 6, 7, 12) | Leadership, Management |
| 13 Pattern 8: The Late Hardware Dilemma (App P) | Management, Stakeholder Rep. |
| 14 Gentle Reminders (Ch 14) | Leadership, Management |
| 15 Take a Challenge (App Q) | Leadership, Management, Stakeholder Rep. |

What should be of greatest value to you is not my specific data, nor my specific observations, but the process which you can use to create your own common scenarios/patterns with your own data. Hopefully you can use your data, in a similar way as I have done in this book, to raise the visibility of the competency needs in your organization leading to prompt and appropriate actions that can help move your organization to the higher competency levels you desire.

**Observations**

The critical competencies I have most often observed a need for, and a shortage of, when it comes to practitioners and their decisions are Leadership, Management and Stakeholder Representation.

This is not meant to imply that I have not observed inadequate design, implementation, testing, and reviews, but rather when these patterns are observed the missing competencies have not turned out to be the ability to code, design, test, or review–but rather to manage lead and communicate.

# Appendix T – Which Organizations Should Care About Essence

The Esssence approach is a proven way to bring agility to the management of your processes, regardless of the type of processes you are currently using and plan to use in the future. This includes organizations that have a mix of project types including both agile and traditional.

Specifically, the approach can help organizations that are currently using traditional life cycles and practices and anticipate continuing the need to support traditional projects, but also want to encourage incremental agile adoption. As another example, the Essence approach can also be of benefit to organizations that want to continue using a separate group to maintain their processes, but at the same time would like to encourage the use of the Essence framework by their practitioners to help them assess the progress and health of their projects and/or would like to encourage the use of the common Essence framework vocabulary.

# References

1 Cockburn, Alistair, Agile Software Development: The Cooperative Game, Second Edition, Addison-Wesley, 2007

2 Chrissis, Mary Beth, Konrad, Mike, Shrum, Sandy, CMMI for Development: Guidelines for Process Integration and Product Improvement, V1.3, Third Edition, Addison-Wesley, 2011

3 Anderson, David, Kanban: Successful Evolutionary Change for Your Technology Business, Blue Hole Press, 2010

4 Poppendeick, Mary, Poppendeick, Tom, Lean Software Development: An Agile Toolkit, Addison-Wesley Professional, 2003

5 Eckes, George, Six Sigma for Everyone, John Wiley, 2003

6 George, Mike, Rowlands, Dave, Kastle, Bill, What is Lean Six Sigma?, McGraw-Hill, 2004

7 Humphrey, Watts, A Discipline for Software Engineering, Addison-Wesley, 1995

8 Humphrey, Watts, TSP: Coaching Development Teams, Addison-Wesley, 2006

9 Glazer, Hillel, CMMI Failure Modes and Solutions – Paving the Path for Agile & CMMI Interoperability

http://prezi.com/vttomwejpe83/cmmi-failure-modes-and-solutions-paving-the-path-for-agile-cmmi-interoperability/

10 Wall Street Journal, Where Process Improvement Projects Often Go Wrong

http://online.wsj.com/news/articles/SB10001424052748703298004574457471313938130

11 Stafford, Jan, What's Behind the Backlash Against Agile?

http://searchsoftwarequality.techtarget.com/feature/Agile-development-Whats-behind-the-backlash-against-Agile?

12 McMahon, Paul, What Should Software Engineering Consist of? An Industry Perspective, Position Paper, 1st Semat Workshop, Zurich, March 2010.

http://www.pemsystems.com/SEMAT_position_McMahon.pdf

(Data originally provided by a Senior Manager in a CMMI Level 5 organization)

13 Reifer, Don, Profiles of CMMI Level 5 Organizations, Crosstalk, January, 2007

http://www.compaid.com/caiinternet/ezine/reifer-profiles.pdf

14 Where Process Improvement Project Go Wrong, Wall Street Journal,

http://online.wsj.com/news/articles/SB10001424052748703298004574457471313938130

15 Avoiding Catastrophic Failures in Process Improvement, Harvard Business Review,

http://blogs.hbr.org/2011/04/avoiding-a-catastrophic-failur/

16 Avoiding Process Improvement Pitfalls,

http://c-spin.net/cspin20080110-AvoidPitfalls.pdf

17 Colvin, Geoff, Talent is Overrated: What Really Separates World Class Performers From Everybody Else, Portfolio, 2008

18 Intentionally left blank

19 Kahneman, Daniel, Thinking Fast and Slow, Farrar, Straus, and Giroux, April, 2013

20 Lehrer, Jonah, How We Decide, Houghton, Miflin, Harcourt, Jan, 2010

21 McMahon, Paul Can I be CMMI Level 5 and Agile Too? (audio), www.pemsystems.com

22 Ambler, Scott, Lines, Mark, Disciplined Agile Delivery: A Practitioner's Guide to Agile Software Delivery in the Enterprise, IBM Press, June, 2012

23 McMahon, Paul, Integrating CMMI and Agile Development: Case Studies and Proven Techniques For Faster Performance Improvement, Addison-Wesley, 2010

24 Buckingham, Marcus, First break all the rules: What the World's Greatest Managers Do Differently, Simon & Schuster, May, 1999

25 The Magical Number Seven Plus or Minus Two

http://en.wikipedia.org/wiki/The_Magical_Number_Seven_Plus_Or_Minus_Two

26 Benson, Jim, DeMaria Barry, Tonianne, Personal Kanban: Mapping Work, Navigating Life, Create Space, 2011

27 Hubbard, Douglas, How to Measure Anything: Finding The Value of Intangibles in Business, John Wiley, 2007

28 Derby, Esther, Agile Retrospectives: Making Good Teams Great, The Pragmatic Programmer, 2006

29 Brooks, Fred, The Mythical Man-month, Addison-Wesley, 1972

30 Putnam, Lawrence, Myers, Ware, Measures For Excellence: Reliable Software On Time, Within Budget, Yourdon Press, 1992

31 McMahon, Paul, Uncommon Techniques for Growing Effective Managers

http://www.pemsystems.com/pdf/Grow_Eff_Mgr_Crosstalk_Update2.pdf,

32 McMahon, Paul, Growing Effective Technical Managers, Presentation at the Systems & Software Technology Conference in Salt Lake City, Utah, 2003

33 Swenson, Keith, Mastering the Unpredictable: How Adaptive Case Management Will Revolutionize the Way That Knowledge Workers Get Things Done, Meghan-Kiffer Press, 2010

34 Kennedy, Michael, Product Development for the Lean Enterprise: Why Toyota's System is Four Times More Productive and How You Can Implement It, The OakLea Press, 2003

35 Liker, Jeffrey, The Toyota Way: 14 Management Principles From the World's Greatest Manufacturer, McGraw-Hill, 2004

36 Intentionally left blank

37 Glazer, Hillel, High Performance Operations, Pearson Education, 2012

38 Gladwell, Malcolm, Blink: The Power of Thinking Without Thinking, Little, Brown and Company, 2005

39 Myburgh, Barry, Towards Understanding the Relationship Between Process Capability and Enterprise Flexibility, Third South African National Conference on software process establishment, assessment and improvement, 2005

40 Patterson, Kerry, Grenny, Joseph, McMillan, Ron, Switzer, Al, Crucial Conversations: Tools for talking when stakes are high, McGraw-Hill 2002

41 Ambler, Scott, Goal Diagrams,

http://disciplinedagiledelivery.com/

42 Lean Six Sigma DMAIC Method, https://en.wikipedia.org/wiki/Six_Sigma.

43 Stratifying Data, http://www.goleansixsigma.com/stratification

44 Stoddard, Robert, Linders, Ben, Sapp, Millee, Exactly What Are Process Performance Models in the CMMI?, http://www.sei.cmu.edu/library/assets/process-models.pdf

45 Hoshin Kanri, http://en.wikipedia.org/wiki/Hoshin_Kanri

46 The British Psychological Society, There is nothing more practical than a good theory

http://www.selfdeterminationtheory.org/SDT/documents/2006_VansteenkisteSheldon_BJCP.pdf

47 Florac, William, Carleton, Anita, Measuring the Software Process: Statistical Process Control for Software Process Improvement, Addison-Wesley, 1999

48 Wheeler, Don, Understanding Variation: The Key to Managing Chaos, Second Edition, SPC Press, 2000

49 Hale, Craig, Rowe, Mike, Do Not Get Out of Control: Achieving Real-time Quality and Performance, Crosstalk, Jan, 2012

http://www.crosstalkonline.org/storage/issue-archives/2012/201201/201201-Hale.pdf

50 Sutherland, Jeff, Jakobsen, Carsten, Scrum and CMMI: Going from Good to Great,

http://jeffsutherland.com/JakobsenScrumCMMIGoingfromGoodtoGreatAgile2009.pdf

51 Ask Why 5 Times, http://en.wikipedia.org/wiki/5_Whys[64]

52 http://en.wikipedia.org/wiki/SEMAT

53 SEMAT website, www.SEMAT.org.

54 OMG Essence Specification, http://www.omg.org/spec/Essence/Current[65].

55 Jacobson, Ivar, Ng Pan-Wei, McMahon, Paul, Spence, Ian, Lidman, Svante, The Essence of Software Engineering: The SEMAT kernel, Oct, 2012, ACMQueue, http://queue.acm.org/detail.cfm?id=2389616

---

[64] http://en.wikipedia.org/wiki/5_Whys

[65] http://www.omg.org/spec/Essence/Current

56 Jacobson, Ivar, Ng Pan-Wei, McMahon, Paul, Spence, Ian, Lidman, Svante, The Essence of Software Engineering: Applying the SEMAT kernel, Addison-Wesley, Jan, 2013

57 Cohn, Mike, Agile Estimating and Planning, Prentice-Hall, 2006

58 Schwaber, Ken, Sutherland, Jeff, The Scrum Guide, The Definitive Guide to Scrum: The Rules of the Game, July, 2011

59 Adkins, Lyssa, Coaching Agile Teams: A Companion For ScrumMasters, Agile Coaches, and Project Managers in Transition, Addison-Wesley, 2010

60 Hypothesis Testing,

http://statstuff.com/index.php?option=com_content&view=article&id=72&vid=sa09

61 Parker, John L, Again to Carthage, Scribner, 2010

62 Parker, John L, Once a Runner, 1978, Scribner Reprint, 2010

# About the Author

Paul E. McMahon (pemcmahon@acm.org), Principal, PEM Systems (www.pemsystems.com) has been an independent consultant since 1997 helping organizations increase agility and process maturity. He has taught software engineering at Binghamton University, conducted workshops on engineering processes and management and has published more than 45 articles and multiple books including *Integrating CMMI and Agile Development: Case Studies and Proven Techniques for Faster Performance Improvement*. Paul is a co-author of *The Essence of Software Engineering: Applying the SEMAT Kernel*. Paul is a Certified Scrum Master and a Certified Lean Six Sigma Black Belt. His insights reflect 24 years of experience working for companies such as Link Simulation and Lockheed Martin, and 17 years of consulting/coaching experience. Paul has been a leader in the SEMAT initiative since its initial meeting in Zurich.

Printed in Great Britain
by Amazon